T0180198

Studies in Systems, Decision and Control

Volume 65

Series editor

Janusz Kacprzyk, Polish Academy of Sciences, Warsaw, Poland
e-mail: kacprzyk@ibspan.waw.pl

About this Series

The series "Studies in Systems, Decision and Control" (SSDC) covers both new developments and advances, as well as the state of the art, in the various areas of broadly perceived systems, decision making and control- quickly, up to date and with a high quality. The intent is to cover the theory, applications, and perspectives on the state of the art and future developments relevant to systems, decision making, control, complex processes and related areas, as embedded in the fields of engineering, computer science, physics, economics, social and life sciences, as well as the paradigms and methodologies behind them. The series contains monographs, textbooks, lecture notes and edited volumes in systems, decision making and control spanning the areas of Cyber-Physical Systems, Autonomous Systems, Sensor Networks, Control Systems, Energy Systems, Automotive Systems, Biological Systems, Vehicular Networking and Connected Vehicles, Aerospace Systems, Automation, Manufacturing, Smart Grids, Nonlinear Systems, Power Systems, Robotics, Social Systems, Economic Systems and other. Of particular value to both the contributors and the readership are the short publication timeframe and the world-wide distribution and exposure which enable both a wide and rapid dissemination of research output.

More information about this series at http://www.springer.com/series/13304

Tian Seng Ng

Real Time Control Engineering

Systems and Automation

 Springer

Tian Seng Ng
School of Mechanical and Aerospace
  Engineering
Nanyang Technological University (NTU)
Singapore
Singapore

ISSN 2198-4182   ISSN 2198-4190   (electronic)
Studies in Systems, Decision and Control
ISBN 978-981-10-9371-5   ISBN 978-981-10-1509-0   (eBook)
DOI 10.1007/978-981-10-1509-0

Printed on acid-free paper

This Springer imprint is published by Springer Nature
The registered company is Springer Science+Business Media Singapore Pte Ltd.

In Memory of
Dr. Lim Tau Meng

Preface

Control technology has been in existence for the last 60 years. Throughout the decades, many developments have evolved in control and automation engineering. Especially, nowadays, when computers have become more popular, technology in control has progress to combine with computer technology for faster and more precise method of computations. Software control can take the place of hardwired control system economically. Hence, new control methods for real-time systems are progressively discovered and taught in institutions. Most engineering applications involve control function. Broadly speaking, there are many types of control technology in engineering domain. We have the electrical and electronics control design, microcontroller and embedded system programmings and control, as well as the mechatronics control system. The low-level assembly programming language performs basic control techniques as well as controlling the stepper motor. Besides, we can find control applications in big and complex industrial system. Power system analysis can predict, monitor and therefore, control the load flow network system. PLC system enhances the design of the elevator control system. We study process control engineering to apply the theory to the water control system. More advanced control technology such as the neural network machine learning technology finds its application in the chemical control plant. Furthermore, computer vision technique is being used widely in the manufacturing factories or the industries. As can be seen, the usefulness of real-time control engineering is applied to countless industrial applications.

The book introduces many different types of control with relevances to real life control systems design. Illustrative diagrams, circuits programming examples and algorithms show the details of the system function design. Readers will find various real-time control automation engineering practices and applications for the modern industries as well as the educational sectors.

Singapore   Tian Seng Ng
May 2016

Acknowledgement

Thanks to the project supervisor, Dr. Fazlur Rahman, for his guidance on the topic: control of process plant using the neural network. Gratefully thanks to Prof. Fok Sai Cheong for his contribution to the control of electro-pneumatic mechanisms. Also, appreciation goes to the technical staff Mr. Lim Chai Lye for assisting in the servo motor control system modification. Besides, the author is also thankful to Mr. Ang Kwee Leng, the engineer for giving his unassuming support on the electrical system.

It was sad that the associate professor, Dr. Lim Tau Meng, had left us. His past contribution to the mechatronics section is highly appreciated. The author would also like to thank Prof. Yap Kian Tiong for his teachings and advises on micro-controller system and Mr. Norbel Navarro for some of the microcontroller pro-grammings. Not forgetting the support team assisting Prof. Yap on the microcontroller section. They are Mdm Yap-Lee Koon Fong, Mrs. Pamela Loh, Mrs. Grace Ho, Mrs. Josephine Loh, Mr. Seow Tzer Fook and Mr. Ng Jui Hock helping in the engineering laboratory.

Appreciation is due to my family members and friends for their overall support and patience. Without these people, the writing of the book would not have been a success. Finally, I would like to thank NTU, School of Mechanical Engineering, Division of Mechatronics and Design, for giving me the opportunity to make this work a reality.

Singapore
May 2016

Tian Seng Ng

Contents

List of Figures

List of Tables

List of Flow Charts

List of Programs

Chapter 1
Introduction

1.1 Objectives

Computer controlled automation, are becoming very common as we apply it in daily life. Advances in technological improvements have made it find its applications in automation. Therefore, the study and understanding of the scope are necessary for building real-time automation systems. More complicated automatic systems require a thorough grasping of the related topics in the book. The materials presented in this book projects the modern control systems and its applications in real time. Readers can gain knowledge in the field of electronics and microcontrollers. The domain provides a good foundation for laboratory-based experiments and practices. Moreover, the readers can also appreciate the electrical knowledge and its applications presented here. Excerpts of the load flow network open an insight to electrical engineers. Interesting computer and programmable control techniques can enhance the interest of the book explorers. We can discover new topics like the neural network automated system and modern technology like computer vision in the book.

1.2 Highlights of the Book

The book introduces many different kinds of control systems and design. They comprise of hardware and software for the real-time engineering control systems.

- Security alarm system is commonly being practiced in almost every organisations. It provides intruder detection during after office hours. We introduced the study of the microprocessor controlled alarm system.
- Electronic circuits were built to accommodate each of the different design structure. We construct various kinds of the mechatronics systems. Some are

© Springer Science+Business Media Singapore 2016
T.S. Ng, *Real Time Control Engineering*,
Studies in Systems, Decision and Control 65,
DOI 10.1007/978-981-10-1509-0_1

programmed using the PBasic programming language. In the book, different types of mechatronics hardware are introduced and appreciated.

- We integrate the microcontroller to perform control and measurements. Numerous types of software are programmed in assembly language to control the hardwired system. We used the 68HC microcontroller family in the system design.
- The design, build and study of the servo motor is found in the book. We experimented the electronic circuitries and analysed the characteristics of the servo motor.
- Programming Logic Control applies in the elevator control design. Ladder diagrams drawn show the flexibility and complexity in the development configuration.
- The complexities of the power system analyses for large system network is being presented in the following chapter. We used the Pascal language in the computations for the power flow calculations.
- Applied control theory is presented to determine the water process control system. Laboratory water tank experiment accelerates the control application.
- The book narrates the control of chemical plant using the neural network for the up to date development in control technology and applications.
- Very often, we applied computer vision technique to the modern industry control system for scanning the defects in the production lines.

1.3   Organisation of the Book

- Chapter 2 analyses the design of the microprocessor security system. We introduce and discuss the hardware circuits and software program in assembly language.
- Chapter 3 describes the many different types of mechatronics control systems. We include diagrams and circuits for the design of the mechatronics systems for better understandings.
- Chapter 4 presents the microcontroller and its hardware applications. We programmed to perform basic i/o functions. Signal conditioning circuits for temperature and pressure measurements are shown. Three different stepper motor control techniques are analysed.
- Chapter 5 introduces the control of a servo motor. Square wave generator is developed for the control input into the servo motor. We present the PID controller in the system. Besides, we also use an analogue controller to control an electro-pneumatic mechanism.
- Chapter 6 highlights the application of the lift control system by using logic ladder diagram. The lift up/down, open/close and light indications functions are illustrated in the complete elevator system.

- Chapter 7 illustrates the power flow techniques for network system analysis. Programming examples in Pascal describes a clear view of how to monitor the load flow system.
- Chapter 8 study the control characteristics of the water tank control system. Experimental set characterises the single tank control system.
- Chapter 9 extends the plant process control by using neural networks. The topic includes the use of the machine learning technique to control the liquid flow of a chemical process plant by controlling the valves.
- Chapter 10 illustrates the use of computer vision. The developed algorithm skeletonized the main outlines of the picture for control and application purposes.

Chapter 2
Embedded Intruder System

A microprocessor is a powerful device used to control the input and output operations of an external device. It serves as a data storage element as well. There are various stages of upgrade and development of the 8086 (16-bit) microprocessor unit. Its applications are used widely in the commercialized and industrialised areas. For example, it is used to perform arithmetic and logical operations inside of a computer system. It enables high speed and large memory storage in computer application, especially in the latest development of the computer system. Besides, CISCO securities also incorporate microprocessor in their security systems.

A standard microprocessor can be programmed and interfaced to the external devices to control and operate a remote control system or collect and perform data operations. An intruder alarm system safeguards an organisation's asset and security. To implement the automation, security system, we need to look at the drawing of the building floor plan and layout. Both the hardware and software combine for the intruder system to work.

The chapter comprises of the detailed design of the microprocessor hardware as well as software algorithms of a security system in the computer room. The scripts described the system analysis, specifications, selection and design of the security system. It also includes hardware interface design and software flow charts and programs. The valuable experience gained and knowledge applied were briefly described in the following few pages.

© Springer Science+Business Media Singapore 2016
T.S. Ng, *Real Time Control Engineering*,
Studies in Systems, Decision and Control 65,
DOI 10.1007/978-981-10-1509-0_2

2.1 Requirements and Assumptions

Hardware Requirements:

- 8086 CPU or its upgraded version
- Microprocessor clock signal of 5 MHz
- 8 KB EEPROM
- 4 KB RAM
- Switches: 5 numbers excluding the reset switch

  - switch 1, 2, 3 for front, side and back door sensors
  - switch 4 to turn on autotime switch
  - switch 5 for external interrupt to alarm

- 7 segment LEDs (8 nos)

  - 6 LEDs to display time in hrs, mins and secs
  - 2 LEDs to display 30 s countdown.

System Analysis:

- At power up system reset and display blank until sw4 is activated.
- Upon pressing the switch, the system starts up and corresponding time of the day is displayed.
- Security system starts from 19:00:00 pm to 07:00:00 am.
- System constantly checked the current time with the limit time.
- If matched, it checks the memory location (60020H) for data.
- After activation, only 30 s are allowed to enter the correct code.
- The system will trigger the alarm if 30 s expire without the correct code entered.
- The system will not trigger an alarm if the correct code is being entered within the time.
- Data will be stored in the memory address (60020H) if someone had entered between the time.
- Otherwise, no data in the location.

Stated Assumptions:

Software:

- Clock generator had no wait state.
- Delay of counting sequence omitted in the program.
- Six secret codes to enter including alphabetical letters within 30 s.
- Assumed program time is aligned to the real-time clock.

Hardware:

- No buffer used for the output device as it is being connected to less than ten output.
- The time colons is by other physical means without using 7-segments.
- External hardware interrupt service provided via interrupt type 60H.
- A reset button (normally closed) is physically connected to the microprocessor via the supply line.
- 4 KB RAM excluding interrupt address.
- Interrupt vector uses another 1 KB RAM.
- 500 byte RAM piece exist.

2.2   Hardware Design

The internal architecture of the CPU consisted of the bus interface unit and the execution unit. They performed the fetch, decode and execution operations. The microprocessor CPU registers enhanced speed processing of data. They are mainly the general purpose registers; pointers and index registers; segment registers and flag registers. The 8086 is used to operate in the minimum mode for single CPU environment in the system. 5 MHz clock operation is designed using 8254A chip and interfaced into the microprocessor. The crystal of the 8254A clock generator is three times the microprocessor input clock. Alternatively, you can use 8284A clock generator to generate a direct 5 MHz timer for the 8086/8088 microprocessor [20].

The selection of the memory or input/output operation enables 1 MB of memory addresses and 64 KB of i/o addresses. The address lines share with the 16 bits data lines. We use latches for latching the address lines in addition to another four address lines. Making a total of 20 address lines. Therefore, it saves the spaces of the microprocessor. Multiplexers are used in the address line so data will not go to the wrong location when we change the address. At the other end, it is demultiplexed. The decoding circuits are used to enable one selection of the addresses at a time. The memory used in the design is 24 KB or 8 KB EEPROM and 2 * 2 KB or 4 KB SRAM. We select static RAM for faster operation due to its non-volatile characteristics. Moreover, we take also the advantages of eliminating the refreshing circuit. The switches connect to the buffer 74LS244 for boosting the signal to the microprocessor. Where latches, 74LS374 are being connected to the seven segment LEDs. The purpose is for stabilizing the output ports. We described the selected design with diagrams (see Figs. 2.1, 2.2 and 2.3).

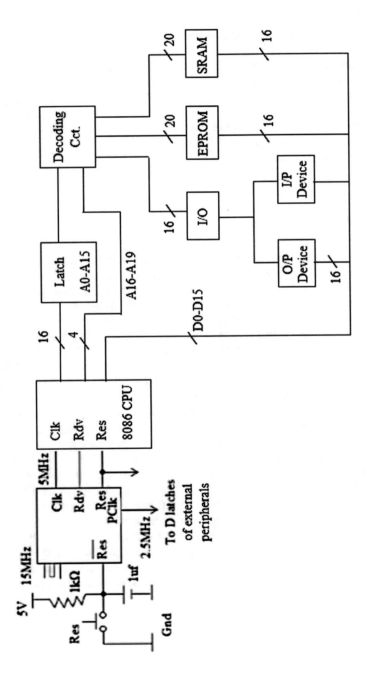

Fig. 2.1 Hardware layout

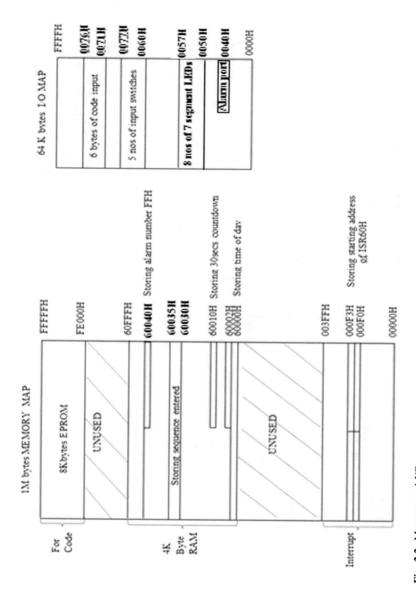

Fig. 2.2 Memory and I/O map

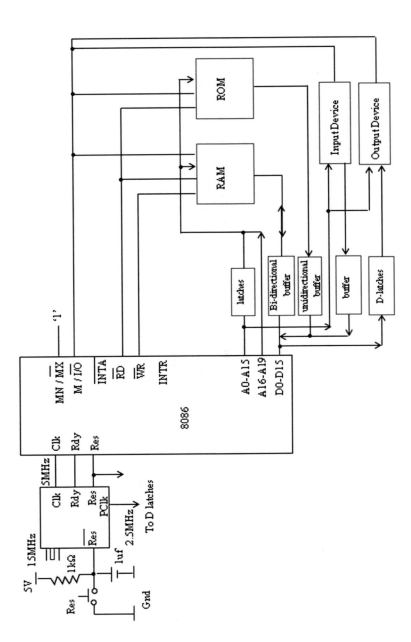

Fig. 2.3 Microprocessor interface

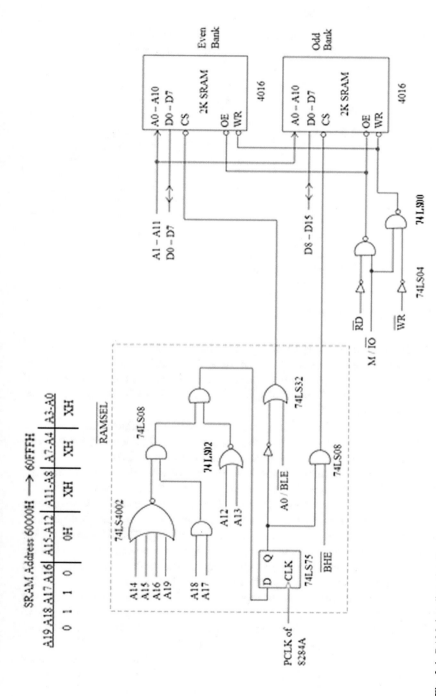

Fig. 2.4 RAM decoding circuit

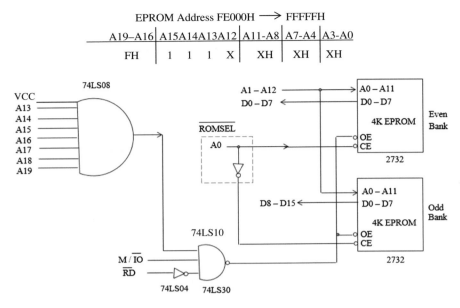

Fig. 2.5 EPROM decoding circuit

In the RAM decoding circuit of Fig. 2.4, the read ($\overline{RD}$) and the write ($\overline{WR}$) operations enable the selections of the data read, and data write operations respectively. We used logic gates to perform the functions of the combinational logics for the addresses. Thus, RAM addresses from 60000H to 60FFFH in the memory map are selectable.

In the EPROM decoding circuit, 2 nos of 74LS08 are required for the 7×2 input AND gates to be joined up for the eight inputs. A0 address forms the necessary selection for the odd or even bank selection (see Fig. 2.5).

Next, we have an alternative solution to the RAM/ROM decoding circuit. The stated assumption in the scenario is that the 1.5 KB and 0.5 KB RAM exists.

The circuit in the top left corner in Fig. 2.6 is derived from the following table. The decoding map represents the 3 rows of memory addresses (see Table 2.1).

Following up is the input decoding device interfacing. The addresses for the input devices range from 0060H to 007EH. The dip switches selections serve as the data as well as the address lines input to the microprocessor unit. The addresses A0 to A4 are selectable by the dip switch input. The relevant dip switch selects the active low for each of the matching input address. For example, each of the dip switches represents each sensor. When any of the five sensors activates, a logic, '0' will select the corresponding address input. So data for the selected address is input. NOR gate helps to decode the address lines A9 to A12. Three of the dip switches are replaceable by door sensors. The input keypad circuit can be designed from address 0071H to 0076H (0000 0000 0111 0xxx) also. Alternatively, we can use 8279 chip for interfacing the keypad to the microprocessor (see Fig. 2.7).

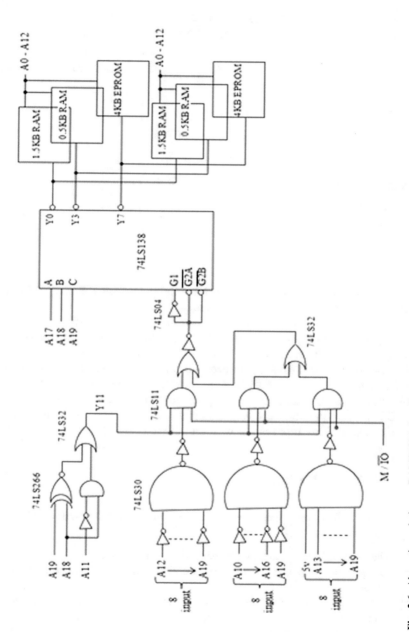

Fig. 2.6 Alternative solution to RAM/ROM decoding

Table 2.1 Decoding table for Y11 output

O/P	C B A A19	A18	A17	A16	A15	A14	A13	A12	A11-A8	A7-A4	A3-A0	Address
Y0	0	0	0	0	0	0	0	0	XH	XH	XH	0H ⟶ 00BFFH (3KB RAM)
Y3	0	1	1	0	0	0	0	0	0 0 X X	XH	XH	60000H → 603FFH (1KB RAM)
Y7	1	1	1	1	1	1	1	X	XH	XH	XH	FE000H →FFFFFH (8KB ROM)

		A11	
C B		0	1
0 0		1	1
0 1		1	0
1 1		1	1
1 0		0	0

$$Y11 = \overline{C}\,\overline{B} + \overline{A11}\,B + C\,B$$
$$= \overline{C}\,\overline{B} + C\,B + \overline{A11}\,B$$
$$= \overline{C \oplus B} + \overline{A11}\,B$$

The output device as shown in Fig. 2.8 is interfaced to the 8 nos. of 7 segment LED display. The seven segment LEDs are for hours, minutes, seconds and the 30 s countdown as stated in the design assumption. We used the BCD to 7 segment driver (74LS47) to light the LEDs. Thus, only the first 4-bit data is enough to display the ten different numbers. The 74LS138 decoder is used to decode the selectable output for display. The decoded output addresses are from 0050H to 0057H. The interrupt type 60H is input into the microprocessor once there is an interrupt request for calling the interrupt subroutine. We can use input port A4 of Fig. 2.7 to call the interrupt request as in Fig. 2.9. The microprocessor will acknowledge the interrupt request to accept the interrupt type.

2.3 Software Design

We include a flow chart, (see Flowchart 2.1) of the program which consists of the following functions:

(a) The auto time will start only when switch 4 activates.
(b) The program consistently scans for the time between 19:00:00 and 07:00:00 interval.
(c) Next, if it is within the time interval, a subroutine will check for the external interrupts. In another word, we scan switch 1 (front door), switch 2 (back door) or switch 3 (side door) for anyone entering the room. If switches are not activated, it will loop back and forth the main program and the subroutine until the time is not within the time interval.

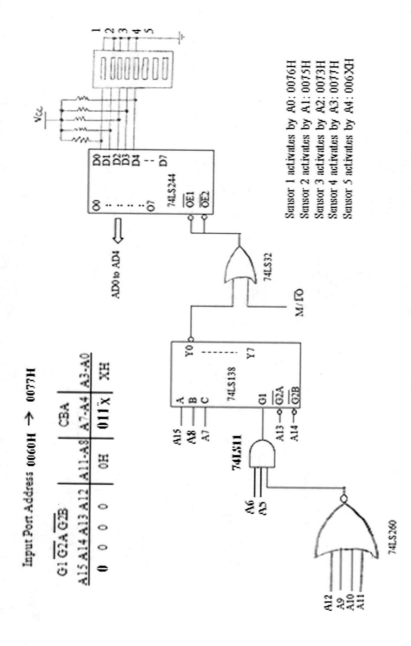

Fig. 2.7 Input device interface. *Note* Dip sw5 activates the input port number 006FH and so on

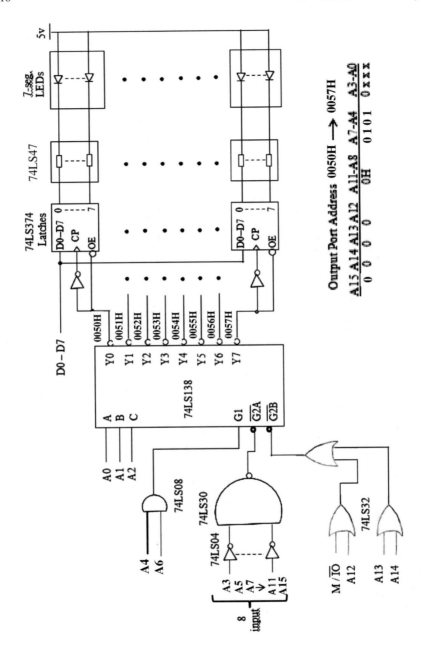

Fig. 2.8 Output device interface

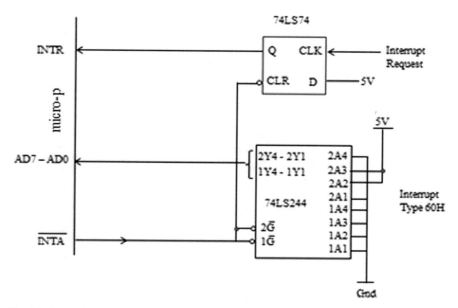

Fig. 2.9 Providing interrupt type to interrupt request

(d) If any one of the switches is activated, it does two things. First, it activates the 30 s countdown. At the same time, it will also check for the correct code sequence entered. The program will check for one input code each time one second passes, by going through the delay procedure. It will compare with the correct code sequence of the six secret codes. Therefore, the program allows the intruder to key in a total of 30 code words before it sounds the alarm. Normally, to get the correct security password take 6 s to key in. But if the user key in wrongly, he may take more than 6 s to do it. If the system detects the correct code within 30 s, it will drop off the alarm.

The range between the first and the last wrong code entered is 54–74 instructions (only consumes 59.2 µs) within the program. If a second passes and no code input, it will take the empty code and compare with the correct code. Since it does not match, the outcome is that it will loop to re-match the code again. Thereby, it lost one second. Similarly, if any of the subsequence code does not match it will also losses a second.

A 16-bit loop will consist of approximately 65,600 maximum instructions. Calculations for each second delay takes about 65,540 instructions per loop. The program consumes a total of 0.052 s per loop (the calculated values is: 4 machine cycles × 200 ns per instruction, for 65,600 instructions). Therefore, the program scans 19 loops for an entered code within a second. So overall, the program may check between 28 and 30 times each time a code is entered, to test for the correct secret code sequence within the 30 s.

Flowchart 2.1 Security
system flow chart

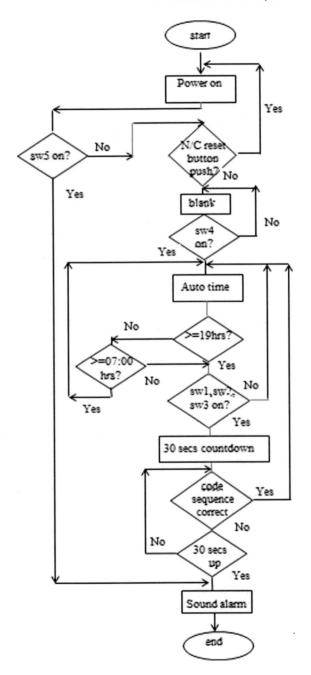

(e) If any of the door sensors is activated and left opened while the code sequence
     enter is correct, the program will continue to run auto time restarting the 30 s
     countdown, to scan for new input codes to be entered. If it senses no input
     entered, it will still sound the alarm if 30 s is up. So, all the doors should be
     shut immediately after entering, between 19.00 pm and 07.00 am. Another
     scenario is, if any of the doors are open and close after entering the correct
     codes, the person will still have to re-enter the codes again.
(f) All the codes entered within the 30 s will be saved.
(g) If 30 s expired before we enter the correct code sequence, the subroutine
     triggers an external interrupt to an alarm.
(h) All the above scenarios will activate the alarm accordingly unless the power
     switch is turned off, or the switch 4 is not on. The alarm is turn off using the
     reset button. The power switch, the reset button and switch 4 is highly secured.

2.4   System Program

We must incorporate hardware with software to run the system properly. An
advantage of the built-in system is that the alarm connected to an interrupt subroutine is triggered externally (see Fig. 2.9) by the hardware (switch 5) for testing
of the alarm system. Secondly, the six code sequence consisting of digits together
with alphabets, which can be displayed by the 7-segment) entered instead of four
digits provides a better security to the system. The stored code entered by the door
breaker can be retrieved by computerized mean by an interface to the microprocessor. Thirdly, the alarm trigger number is stored in the memory location. It serves
to check for intruder break in for the case where the alarm fails to sound.

A disadvantage of the system is that it affects the auto timing when the 30 s
activates and the time taken to enter the code sequence. Thereby, causing a delay in
the actual timing of the system. It is considered as pros, as the delay in the timing
system might be a second alternative to hint for a broke-in between the time
interval, besides checking the code sequence stored. Later the system can be reset
back to the precise auto time again by the personnel, which have the key to the reset
button panel. We can also improve the system further.

Program 2.1:Microprocessor Security System

```
;----------------------------------------------------------------------
;DATA SEGMENT DEFINATION
;----------------------------------------------------------------------
DATA_S  SEGMENT  PUBLIC  'DATA'   ; HAS TO BE IN RAM AREA
        ORG   60000H
        TIME  DB  ?           ; SECONDS COUNTER
              DB  ?           ; MINUTES COUNTER
              DB  ?           ; HOURS COUNTER (24 HOURS CLOCK)
        TEMP  DB  3  DUP(0) ;TEMPORARY STORAGE FOR SECS;MINS;HRS
     TYPE_60  DW  2  DUP(0) ; STORAGE FOR ALARM INTERRUPT ISR
  DELAYCNT DB  32H  ;DELAY FOR 50 TIMES
;----------------------------------------------------------------------
;SEVEN SEGMENT REPRESENTATIVES
;----------------------------------------------------------------------
                    ; '0'   '1'   '2'   '3'   '4'   '5'   '6'   '7'   '8'
     SEVEN_SEG  DB  3FH,  06H,  5BH,  4FH,  66H,  6DH,  7DH,  07H,  7FH
                    ; '9'   'A'   'B'   'C'   'D'   'E'   'F'
                DB  6FH,  77H,  7CH,  39H ,  5EH,  79H,  71H
;----------------------------------------------------------------------
;THIRTY SEGMENT REPRESENTATIVES
;----------------------------------------------------------------------
                     ; '0'   '1'   '2'   '3'   '4'   '5'   '6'   '7'   '8'
     SIXTY_SEG  DB  00H,  01H,  02H,  03H,  04H,  05H,  06H,  07H,  08H
                DB  09H,  10H,  11H,  12H ,  13H,  14H,  15H,  16H,  17H
                DB  18H,  19H,  20H,  21H ,  22H,  23H,  24H,  25H,  26H
                DB  27H,  28H,  29H,  30H,  31H,  32H,  33H,  34H,  35H
                DB  36H,  37H,  38H,  39H ,  40H,  41H,  42H,  43H,  44H
                DB  45H,  46H,  47H,  48H,  49H,  50H,  51H,  52H,  53H
                DB  54H,  55H,  56H,  57H,  58H,  59H,  60H
;----------------------------------------------------------------------
;SECRET CODE (TO OFF ALARM)
;----------------------------------------------------------------------
D8E1A1       DB       E8H              ; 'D'
             DB       D7H              ; '8'
             DB       EDH              ; 'E'
             DB       78H              ; '1'
             DB       DDH              ; 'A'
             DB       78H              ; '1'
;----------------------------------------------------------------------
;LOOK-UP TABLE FOR CODE IN ASCII (TO OFF ALARM)
;----------------------------------------------------------------------
```

```
                       ; '0'   '1'   '2'   '3'   '4'   '5'   '6'   '7'   '8'   '9'
;D8E1A1:     DB    77H, 78H, 7DH, 7EH, B7H, BBH, BDH, BEH, D7H, DBH
             ;DB   DDH, DEH, E7H, E8H, EDH, EEH
DATA_S ENDS        ; 'A'   'B'   'C'   'D'   'E'   'F'
```

```
;----------------------------------------------------------------------------
;EXTRA SEGMENT DEFINITION
;----------------------------------------------------------------------------
EXTRA_S  SEGMENT  PUBLIC  'EXTRA'  ; HAS TO BE IN RAM AREA
         ORG        60030H
         CODESEQ DB 6 DUP(?)  ; STORAGE OF CODE ENTERED
EXTRA_S  ENDS
```

```
;----------------------------------------------------------------------------
;STACK SEGMENT DEFINITION
;----------------------------------------------------------------------------
STACK_S  SEGMENT  STACK  'STACK'  ; HAS TO BE IN RAM AREA
             DW   512     DUP(?)
             TOS  LABEL  WORD
STACK_S  ENDS
```

```
;----------------------------------------------------------------------------
;CODE SEGMENT DEFINITION
;----------------------------------------------------------------------------
CODE_S  SEGMENT  PUBLIC  'KEY' ;CODE-SEG STARTS AT ROM FE000H
            ASSUME    CS:CODE_S, DS:DATA_S, SS:STACK_S, ES:EXTRA_S

            MOV TYPE_60+2, SEG ISR60H   ;CS LOCATION FOR ALARM ISR
            MOV TYPE_60, OFFSET ISR60H  ;IP OF ALARM ISR
```

```
;----------------------------------------------------------------------------
;INPUT (SWITCH) PORT ADDRESS DEFINITION
;----------------------------------------------------------------------------
      SW1    EQU    0076H        ; INPUT PORT ADDRESS FOR SWITCH 1
      SW2    EQU    0075H        ; INPUT PORT ADDRESS FOR SWITCH 2
      SW3    EQU    0073H        ; INPUT PORT ADDRESS FOR SWITCH 3
      SW4    EQU    0077H        ; INPUT PORT ADDRESS FOR SWITCH 4
      SW5    EQU    006FH        ; INPUT PORT ADDRESS FOR SWITCH 5
```

```
;----------------------------------------------------------------------------
;INPUT (CODE) PORT ADDRESS DEFINITION
;----------------------------------------------------------------------------
            CODE1 EQU  0071H        ; INPUT PORT FOR FIRST CODE
            CODE2 EQU  0072H        ; INPUT PORT FOR SECOND CODE
            CODE3 EQU  0073H        ; INPUT PORT FOR THIRD CODE
```

```
              CODE4  EQU   0074H      ; INPUT PORT FOR FOURTH CODE
              CODE5  EQU   0075H      ; INPUT PORT FOR FIFTH CODE
              CODE6  EQU   0076H      ; INPUT PORT FOR SIXTH CODE
```

```
;-------------------------------------------------------------------------------------
;OUTPUT (7 SEG DISPLAY) PORT ADDRESS DEFINITION
;-------------------------------------------------------------------------------------
         OUTS1  EQU   0050H  ; OUTPUT PORT ADDRESS FOR SECOND 1
         OUTS2  EQU   0051H  ; OUTPUT PORT ADDRESS FOR SECOND 2
         OUTM1  EQU   0052H  ; OUTPUT PORT ADDRESS FOR MINUTE 1
         OUTM2  EQU   0053H  ; OUTPUT PORT ADDRESS FOR MINUTE 2
         OUTH1  EQU   0054H  ; OUTPUT PORT ADDRESS FOR HOUR 1
         OUTH2  EQU   0055H  ; OUTPUT PORT ADDRESS FOR HOUR 2
         OUTC1  EQU   0056H  ; PORT ADDRESS FOR low byte of 30 secs
         OUTC2  EQU   0057H  ; PORT ADDRESS FOR high byte of 30 secs
```

```
;-------------------------------------------------------------------------------------
;============ MAIN PROGRAM ===========================================
;-------------------------------------------------------------------------------------
MAIN       PROC        NEAR
START:     CLI
           MOV    AX,  DATA_S      ; INITIALISATION
           MOV    DS,  AX
           MOV    ES,  AX
           MOV    AX,  STACK_S
           MOV    SS,  AX
           MOV    SP,  OFFSET  TOS
ON:        MOV    AX,  00H
           IN     AX,  SW4         ; check port switch 4
           AND    AL,  00001000b   ; check autotime on
           CMP    AL,  08H
           JNE    CLOCK            ; starts autotime if sw 4 is on
           JMP    ON
;============ CLOCK TIME ( 8 NOS 7 SEGMENT) DISPLAY ============
CLOCK:  CALL  TIMES
           MOV    SI,  0H
           MOV    DX,  0H
           MOV    BL,  0H
           MOV    AX,  0H
           MOV    CX,  03H                 ; SET TO LOOP 3 DB FOR HR-MIN-SEC
ONE:     STOSB TEMP[SI], TIME[SI];STORAGE FOR THE HR-MIN-SEC TIME
           MOV    BX,  OFFSET  SIXTY_SEG
RET:     MOV    AL,  TEMP[SI]
           XLATB
           AND   AL,  0FH
```

```
          MOV    DH,  50H+DL
          ADD    DH,  CH
          OUT    DH,  AL          ; EVEN 7-SEGMENT TIME DISPLAY
          MOV    AL,  TEMP[SI]
          XLATB
          AND    AL,  F0H
          MOV    DH,  51H+DL
          ADD    DH,  CH
          OUT    DH,  AL          ; ODD 7-SEGMENT TIME DISPLAY
          DEC    CL
          JZ     HERE
          INC    SI
          INC    CH
          INC    DL
          LOOP   ONE
HERE:     MOV    AL,  13H; MOVE 19HRS IN HEX INTO AL
          CMP    AL,  TEMP[SI] ; CHECK FOR 19 HOURS
          JGE    GOTO   ; CHECK DOOR SENSORS IF 19HRS AND ABOVE
          MOV    AL,  07H        ; MOVE 07HRS IN HEX INTO AL
          CMP    AL,  TEMP[SI] ; COMPARE WITH REAL-TIME HRS
          JGE    CLOCK    ; CONTINUE TIMING CLOCK IF 07HRS & ABOVE
GOTO:     CALL   CHKSENSOR
          JMP    CLOCK          ; CONTINOUS TIME DISPLAY
MAIN      ENDP
```

```
;----------------------------------------------------------------------------------------
;=========== TIME AUTO ADJUSTMENT PROCEDURES ================
;----------------------------------------------------------------------------------------
TIMES    PROC   NEAR
                PUSH   AX      ; save registers
                PUSH   BX
                PUSH   SI
                MOV    AH,  3CH        ; load modulus 60 of counter
                MOV    SI,  BX
                MOV    BX,  OFFSET  TIME ; address time
                CALL   DELAY
                CALL   UP              ; adjust seconds counter
                MOV    AH,  AH
                JNZ    DONE
                INC    SI
                MOV    AH,  3CH
                CALL   UP              ; adjust minutes
                MOV    AH,  AH
                JNZ    DONE
                INC    SI
```

```
                        MOV    AH, 18H    ; modulus 24
                        CALL   UP
            DONE:       POP    SI                    ; restore registers
                        POP    BX
                        POP    AX
                        RET
TIMES   ENDP
UP      PROC            NEAR
                        MOV    AL, [SI]
                        ADD    AL, 1          ; increment counter
                        DAA                   ; keep in BCD format
                        MOV    [SI], AL
                        MOV    AL, AL         ; change to hex format
                        SUB    AH, AL
                        JNZ    UP1
                        MOV    [SI], 0H
            UP1:        RET
UP      ENDP


;----------------------------------------------------------------------------------------
; == PROCEDURE CHECK FOR FRONTDOOR, SIDEDOOR & BACKDOOR OF
;== SENSOR1, SENSOR2 & SENSOR3 CORRESPONDINGLY FOR OPEN-CCT
; == (NOSIGNAL OR DOOR OPENED) ===================================
;----------------------------------------------------------------------------------------
CHKSENSOR       PROC    NEAR
                        PUSH   AX
                        PUSH   BX
                        PUSH   CX
            CHECKS:     MOV    CL, 03H
                        MOV    BL, 00H
            ON:         MOV    AL, 0H
                        IN     AL, [ SW1 +BL]  ; check 3 sensors
                        CMP    AL, 76H
                        JE     LINK                  ; if detected
                        CMP    AL, 75H
                        JE     LINK
                        CMP    AL, 73H
                        JE     LINK
                        DEC    CL
                        JZ     RETURN
                        INC    BL
                        LOOP   ON
                        JMP    RETURN
            LINK:       CALL   COUNTDOWN    ; if door opened, starts countdown
            RETURN:     POP    CX
```

```
                        POP      BX
                        POP      AX
                        RET
CHKSENSOR               ENDP ;
```

;--
;============= OUTPUT ALARM INTERRUPT =========================
; ---

```
ISR60H        PROC    NEAR
              MOV     DL,    0040H   ; ALARM PORT LOCATION
              MOV     AX,    FFH
              MOV     60040H,  AX    ; STORE ALARM NUMBER
              OUT     DL,    AX
              IRET
ISR60H        ENDP
```

;--
;============= DELAY OF ONE SECOND PROCEDURES =============
; ---

```
DELAY        PROC   NEAR
             MOV    BL,    DELAYCNT   ; SETUP DELAY DURATION
             MOV    BH,    00H
LOOP1:       MOV    CX,    16EAH        ; 20msecs DELAY
             DEC    BX
LOOP2:       LOOP   LOOP2
             JNZ    LOOP1
             RET
DELAY        ENDP
CODE_S       ENDS
```

;--
; ============= 30 SECONDS COUNTDOWN PROCEDURE =============
;============= & CHECK FOR CORRECT CODE SEQUENCE ==========
;--

```
ALARM_CHECK    SEGMENT
COUNTDOWN      PROC    FAR
ASSUME      CS:ALARM_CHECK,  DS:DATA_S
             PUSH    AX
             PUSH    BX
             PUSH    CX
             PUSH    DX
             PUSH    SI
             PUSH    DI
COUNTS:      MOV     BX, 0H
             MOV     CX, 0H
             MOV     DX, 0H
             MOV     AL, 00H
```

```
                MOV     SI,  0H
                MOV     DI,  0H
                MOV     SI,  OFFSET D8E1A1    ; OFFSET CORRECT CODE
                MOV     DI,  OFFSET  CODESEQ     ; OFFSET ENTERED CODE
                MOV     BX,  OFFSET  SIXTY_SEG
                MOV     CL,  0H
                MOV     CH,  06H
SEQUENCE:       MOV     AL,  1EH      ;30 DEC OR SECONDS
                SUB     AL,  CL
                XLATB
                AND     AL,  0FH
                OUT     OUTC1,  AL ; DISPLAY LOW BYTE OF 30 SECS CNTDN
                MOV     AL,  1EH
                SUB     AL,  CL
                XLATB
                AND     AL,  F0H
                OUT     OUTC2,  AL ; DISPLAY HIGH BYTE OF 30 SECS CNTDN
                JMP     DELAY       ; DELAY OF 1 SEC BEFORE ENTER NEXT
                                    ; CODE
                MOV     AL,  0H
                MOV     DX,  CODE1
                ADD     DX,  SI
                IN      AL,  [DX] ; ENTER CODE
                STOSB   CODESEQ[DI],  AL ; STORE IN EXTRA SEG. 60030H
                INC     CL
                CMP     CL,  1EH    ; COMPARE WITH 30 SECS
                JE      ALARM    ; 30 SECS UP
COMPARE:        CMP     CODESEQ[DI],   D8E1A1[SI]
                JNE     SEQUENCE         ; CHECK FOR CORRECT CODE
                DEC     CH               ; IF CORRECT, GOTO NEXT CODE
                JZ      STOP             ; IF 6 CORRECT CODE, STOP
                INC     SI
                INC     DI
                JMP     SEQUENCE         ; FIND NEXT CORRECT CODE
ALARM:          INT     60H              ; 30 SECS UP TRIGGERS INTERRUPT
STOP:           POP     DI               ; REQUEST TO ACTIVATE ALARM
                POP     SI
                POP     DX
                POP     CX
                POP     BX
                POP     AX
                RET
COUNTDOWN       ENDP
ALARM_CHECK     ENDS
END
```

Chapter 3
Mechatronics

Mechatronics systems are widely utilized in the automation, industries. The practices involve the controls of electronics for mechanical systems. In this chapter, we present the liquid level control system, the oscillating planar mechanical system, the conveyor inspection using shift registers and the modern speed control unit.

3.1 Liquid Level Control

The latching circuit for filling the water into a tank or beaker is as shown in Fig. 3.2. We can conduct a simple experiment by using a solenoid valve to open and close an opening for controlling the flow of water into a beaker. The operation is simply a latched circuit for water flowing to a beaker. The apparatus requires a water carrying container hooked onto a standing rod structure and a beaker. The bar structure only has a square base and an upright standing bar or rod. The square base holds the weight of the water container. We placed a beaker underneath the water carrying container, on the base frame structure. A solenoid is attached below to the container to open or close the valve. The solenoid valve is at normally closed when not energise, to stop the water from flowing out of the container. A small pipe fixed downward from the container leads the water to fill up the beaker.

The sensors required are a proximity sensor, a latch button and a spring button. The latch button 'S2' is latched for the operation of the system. The operation starts by pressing the 'LS1' spring button. The output of the 'OR' gate is activated. The LED is lighted up by the logic '1' from the 'OR' gate output. The signal also

© Springer Science+Business Media Singapore 2016
T.S. Ng, *Real Time Control Engineering*,
Studies in Systems, Decision and Control 65,
DOI 10.1007/978-981-10-1509-0_3

Fig. 3.1 Transistor TIP-121

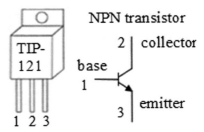

enters the buffer to energise the solenoid valve. Thus, the valve will open and water start to flow out from the container into the beaker. The output of the 'OR' gate also goes into the Pin 1 of the 'AND' gate. So the circuit is latched as the 'AND' gate outputs a logic '1' signal. In the circuit, the 1 kΩ resistors are being used as the pull-down resistors. We connect the 1.8 kΩ resistors to the transistor base. The 470 Ω resistor serves as the pull-up resistor for the proximity sensor.

We placed the proximity switch at a certain level of the beaker. So that when the water level reaches the level, the proximity switch will be activated to stop the water flow. The 'TIP-121' transistor will be de-activated and a logic '0' will go to the pin 2 of the 'AND' gate. Thus, the output of the 'AND' gate will be zero. The output LED will not lit up, and the solenoid valve will be de-energised. So the opening of the valve will close back to stop the water from flowing out of the pipe. If the latched button is de-latched during the water filling operation, it will stop the water filling process. To continue the operation, we had to reset the latch switch to the connected position. Then we pressed the 'LS1' spring button to continue the water filling process (Fig. 3.1).

An illustration of a simple program written in the PBasic language is as shown in the program P3.1. We make use of the Basic Stamp microcontroller to perform similar function. The microcontroller connects to three input ports 5, 6 and 7 for sensings. Input port 7 connects to the proximity sensor. While we connect the input ports 5 and 6 to LS2 and S2 (latched) switches respectively. If both buttons are turned on, it will activate the solenoid to open the valve to fill the beaker or water tank. The proximity sensor (to IN7) fixed at a certain level of the tank must also be cleared to begin the water flowing process. The output port 10 is connected to activate the solenoid valve via a buffer and a TIP-121 transistor. Similarly, a 1.8 kΩ resistor is connected to drive the base of the transistor. The proximity sensor is activated when the water level reaches its height. Thereby

```
--------------------------------------------
Program 3.1: Liquid Level Control
--------------------------------------------
'{$STAMP BS2}
DIRS=%0000000010000000
again IF IN7=1 AND IN5=1 THEN do
  you:OUT10=0 ' water stop
      GOTO again
  do: IF IN6=0 OR IN7=0 THEN you
      OUT10=1 ' water flow
      GOTO do
```

causing IN7 to input a zero, resulting output port 10 to output a logic low to the transistor. It will deactivate the solenoid, and close the valve to stop the water from flowing out. By making use of the program, we can replace the hardware circuits (Fig. 3.3).

3.2 Oscillating Planar

In this section, we combined the field of electronics with mechanical to build and control an oscillating planar. The operation begins with the preset switch pressed when powered up. The first D-latch on top presets with a logic '0' and the second D-latch below clears with a logic '0'. So the oscillating planar starts to turn to the right. As the planar rotates, the slotted optical switch will be blocked and activates again. It triggers clock input to the D-latches. As a result, a logic '0' now output from the top D-latch. The logic '1' now appear at the output of the lower D-latch. It reverses the turning direction of the oscillating planar. It will again activate both the optical switches fixed on both sides of the planar's edge. A second trigger to the D-latches will stop the planar. Now both the outputs of the D-latches are zero. Therefore, we can see that the function of the preset switch is to reset the position of the oscillating planar to its starting position. The circuit is operated by a start switch to hold a logic '1' to pin 2 of the 'AND' gate. We begin the oscillating function by pressing the preset switch. It triggers a logic high and a logic low from the output of

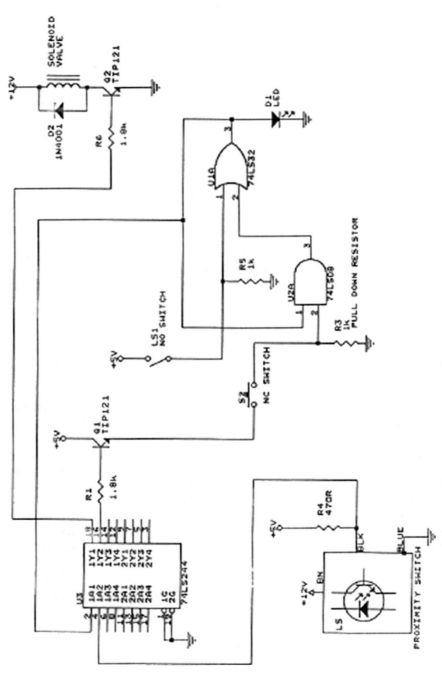

Fig. 3.2 Latching circuit for filling water tank

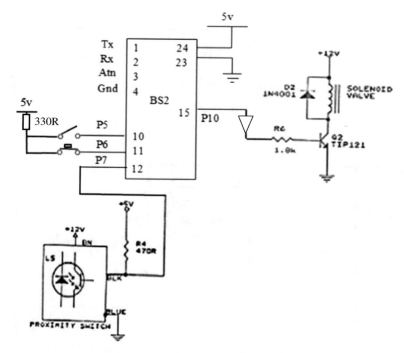

Fig. 3.3 Microcontroller controlled latching circuit

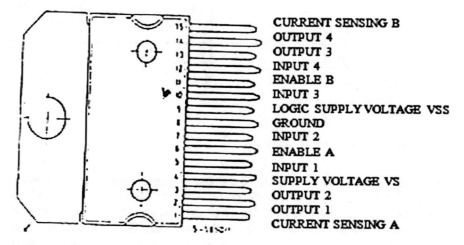

Fig. 3.4 L298 H-bridge dual bidirectional motor driver

Table 3.1 L298 logic
control

Input		Function
VA = H	C = H; D = L	Turn right
	C = L; D = H	Turn left
	C = D	Fast motor stop
VA = L	C = X; D = C	Free running motor stop

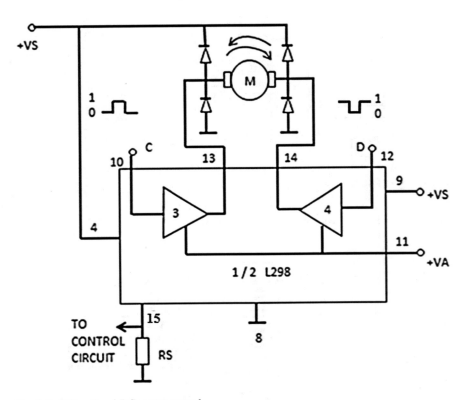

Fig. 3.5 Bidirectional DC motor control

the D-latches to turn the planar. As the slotted optical switches senses, a clock trigger to the D-latches will output a logic '0' to the 'C' input of the motor. At the same time logic '1' enters the 'D' input of the same motor simultaneously. It changes the direction of the moving planar. The oscillating planar reverses direction again when the optical switches activate a second time. The operation continues until the 'start' holding switch stops. So we can see that we use the same button for the start and the stop operation (Fig. 3.4 and Table 3.1).

The 4 diodes used for the DC motor are 1 A high speed diodes (Figs. 3.5, 3.6).

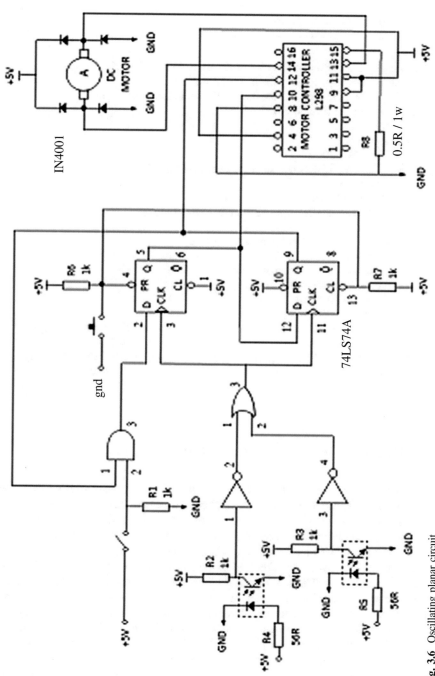

Fig. 3.6 Oscillating planar circuit

Program 3.2: Oscillating Planar

```
'Oscillating Table Solenoid Sense:Software to replace flip-flop control
'in2=clear; in5=reject; in6=sensor; out12=red led
'out13=yellow led; out14=green led; out15=solenoid+orange led
DIRS=%1111111100000000
  cnt VAR Byte
  gone: cnt=0
        OUTD=%0000
  clear: IF IN5=1 THEN red 'sense reject push button
        GOTO clear
  check:IF IN6=0 THEN clk  'sense sensor
        IF IN2=0 THEN gone 'sense clear push button
        GOTO check
  clk: IF IN6=1 THEN jump 'pulse off sensor
        GOTO clk
  jump: cnt=cnt+1        'count number of sensing
      IF cnt=4 THEN gone
      IF cnt=3 THEN out
      IF cnt=2 THEN green
        OUTD=%0010 'yellow            green: OUTD=%0100 'green
        GOTO check                          GOTO check
  out: OUTD=%1000 'orange &          red: OUTD=%0001   'red
        GOTO check 'solenoid                GOTO check
```

3.3   Conveyor Inspection Using Shift Registers

The circuit implements shift registers to control a diverter solenoid. Four negative edge-triggered J-K flip flops as well as a slotted optoisolator, are the main electronics components for the circuit. The far end of the conveyor is where we located the diverter. We mount small pieces of block indicator at a fixed distance along the conveyor system. The block indicator will energize the slotted switch at distance intervals as the conveyor rolls. The start of the conveyor and the diverter end is three block indicators away. There is indicator block at the starting point of the conveyor system where we placed the material. We used a reject switch or a proximity sensor to detect material put on the conveyor system. A proximity sensor is mounted at the beginning of the conveyor system where we placed the material. The output of the first J-K flip-flop will light up if something is on the conveyor system. As the conveyor rolls, the J-K flip-flops will be activated when the indicator block covers the slotted optoisolator. It will lit up the second LED at the output of the second J-K

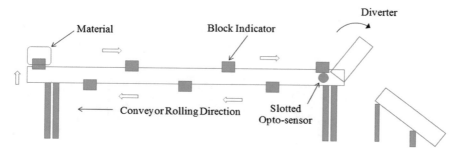

Fig. 3.7 Conveyor system

flip-flop if a material is present. When the material travels to the end of the diverter, the output of the fourth J-K flip-flop will be energised. At the same time, the diverter solenoid will be energized to divert the material placed on the conveyor system. The LEDs indicate which zone the material has arrived in the midst of travelling in the conveying system. If we put two materials subsequently in each different indicator block, two LEDs will light up each time the indicator blocks energize the slotted optical sensor. When nothing is on the conveyor, the far end diverter will not be activated as the last J-K flip-flop outputs a zero (Figs. 3.7, 3.8).

3.4   Modern Speed Control

We can create a small permanent magnet motor control system by pulse-width modulation technique. Two 555 timer [16] electronic components are required for the circuitry. The circuit design is as shown in Fig. 3.9 Pulse-width modulation control system for a small permanent magnet motor. We supply both the timers with a +15 V DC source. We connect the first timer on the left as a square wave oscillating circuit while the second timer on the right connects as a one-shot. The RC differentiator circuit contains the resistor Rdiff and the capacitor Cdiff. The first 555 timer of the circuit determines the rise and fall time of the deterministic output signal. Each time the output from the first timer goes low, the differentiator delivers a fast negative going spike to trigger the input of the second 555 timer, with the spike bottoms at 0 V. That initiates another pulse output from the second 555 timer. The period for the second timer input is approximately 25 ms. The operation of the RC differentiator circuit starts with a low voltage to the negative going side of the Cdiff. The current flows to charge up capacitor Cdiff and reduces input flow to the timer. It triggers a negative spike to the input of the second timer. When we charged the Cdiff to full capacity, the input to the second timer is at 0 V. When the output of the first timer inverts back from the low level output the capacitor Cdiff discharges. All current flows from Cdiff to the trigger input of the second timer.

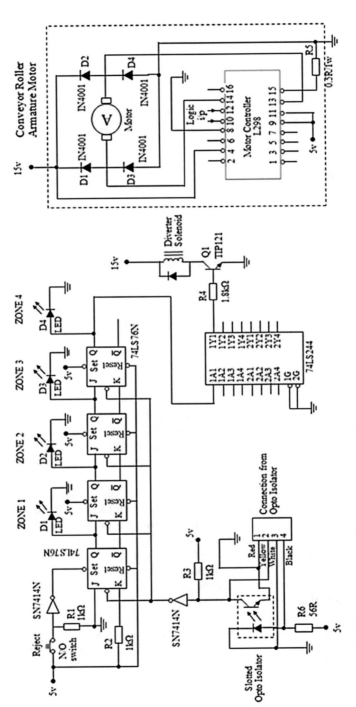

Fig. 3.8 Shift register conveyor control

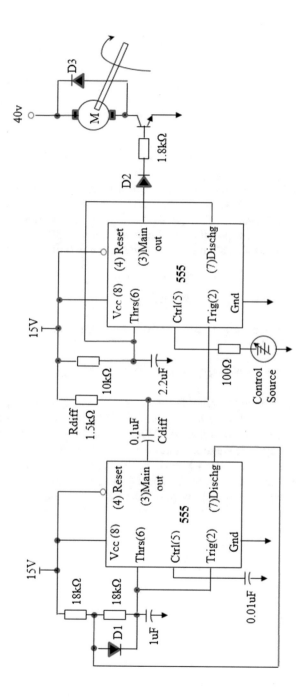

Fig. 3.9 PWM control of a small permanent magnet motor

Fig. 3.10 RC differentiator
pulse signal

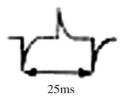

25ms

It causes a positive spike to the trigger input of the timer. The rate of discharge depends on Cdiff and Rdiff (Figs. 3.9, 3.10).

The RC differentiator circuit time constant is 0.15 ms as we used a 1.5 kΩ and the 0.1 uF passive components. We determined the period by using resistors and capacitor of 18 kΩ and 1 uF respectively. By the connection of the 0.01 uF capacitor to the control input, the 555 timer oscillates at 50 % duty cycle, with a period of 25 ms. The calculation for the timing of the first timer circuit is as follow.

$$T = 2 \times 0.7 \times 18 \text{ k} \times 1\text{u}$$
$$= 25.2 \text{ ms.}$$
(3.1)

We connected the resistor and capacitor from the second timer as 10 kΩ and 2.2 uF respectively.

$$T = RC = 10 \text{ k} \times 2.2 \text{ u} = 22 \text{ ms.}$$
(3.2)

As can be seen, it forms the variable time peak of 22 ms. Therefore, the maximum duty cycle of the second timer is reachable at (22 ms/25 ms) 88 % efficiency. The 40 v permanent magnet motor is controllable with a maximum duty cycle of 88 %. The motor's duty cycle is variable from 10 to 88 % by varying the control source to the second one-shot timer from 1.8 to 10 V. We control the mechanical shaft power of the permanent magnet motor by the control source. A +40 V DC source powered the permanent magnet motor. Low pass filter can be installed to the motor when needed.

Chapter 4
Microcontroller

Microcontrollers require the digital signal to perform functions and applications. For ensuring the analog input signals matched that of the digital input of the microcontroller, the analog signal must be within the range of the digital device with TTL or CMOS logic level compatibility. Moreover, the output impedance of the sending device should be small enough for the transfer of power to the receiving device. Conversely, the input device should have a high input impedance to match the output device. It is related to the driving current specifications for the output of the sending devices and the sinking current of the input receivers.

4.1 Basic I/O Modules

Load and run program 4.1. Note that port A connector is internally being connected to 10 kΩ pull-up resistors. Observe how to read in a 8 bits information via Port A. Note that the address of Port A is $1000. DDRA address at $1001. The register controls whether the bit 0–7 use as input (Low) or output (High) pins. Index addressing with register X. CLI (clear interrupt mask) to enable interrupts by setting I bit to zero. By doing so, we enable the software debugger. Check the value at address $0050 using the debugger. It should give you the value indicated by the dip switches. All files associate with the file extension .a11 in the assembly format (Fig. 4.1).

© Springer Science+Business Media Singapore 2016
T.S. Ng, *Real Time Control Engineering*,
Studies in Systems, Decision and Control 65,
DOI 10.1007/978-981-10-1509-0_4

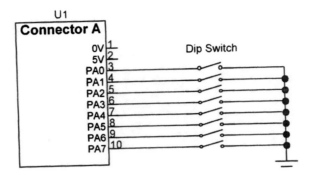

Fig. 4.1 8-bit dip switch input circuit

Program 4.1: Read Dip Switch

```
porta  equ  0
ddra   equ  1
       org  $8000          ; Micro-P starting address
       cli                 ; enable the debugger
       ldx  #$1000
       ldaa #$00
       staa ddra, x  ;set poara as input
again  ldaa porta, x ;load dip switch status
       staa $0050          ;store dip switch status
       bra  again
```

Program 4.2: Write to Output LEDs

```
portb equ $60                    staa  portb, x
      org $8000                  jsr  delay
      cli                        bra  again
      ldx #$1000                 org  $8200
again ldaa #$0f            delay ldy  #$ffff
      staa portb, x        loop  dey  #$1
      jsr delay                  bne  loop
      ldaa #$f0                  brts
```

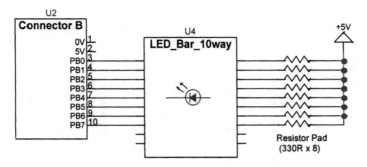

Fig. 4.2 LEDs output circuit

Note that port B (Fig. 4.2) of the output circuit is connected internally to ULN2803A Darlington transistors. You can load and run the program 4.2 to observe the following points. How to output a 8 bits output via Port B. The address of Port B is $1060. That is it is Port B of the basic microprocessor, not the Port B of the 68HC11 microcontroller. We observe the branching and looping techniques.

```
----------------------------------------------
Program 4.3: Push Button & Buzzer
----------------------------------------------

portd  equ  8
ddrd   equ  9

           org  $8000                 ; Micro-P starting address
           cli                        ; enable the debugger
           ldx  #$1000
           ldaa #%00100000            ;set pd2 as input & pd5 as output
           staa ddrd, x
begin  bclr portd, x $20              ;clear bit pd5
again  brset portd, x $04 begin       ;check bit pd2
           bset portd, x $20          ;set bit pd5
           bra again
```

We connect connector D to port D and port G (Fig. 4.3). The function of program 4.3 is to activate a buzzer using a push button. Make sure you understand the following points. Address of port D is $1008. Address of Data Direction Register for port D (DDRD) is $1009. This register controls whether bit 0–5 use as input (LOW) or output (HIGH) pins. Bit 6 and 7 are not usable. Index addressing

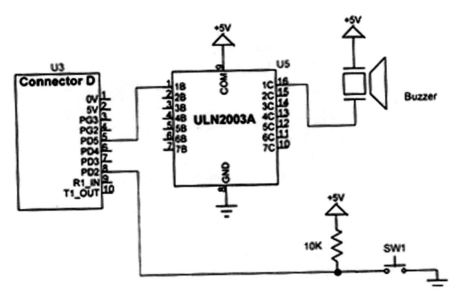

Fig. 4.3 Buzzer and push button circuit

with register X. We used branching and looping. The high current Darlington transistor arrays are for driving the buzzer load.

In basic, both programs 4.4 and 4.5 are the same. They read in the PE2 port for the A/D conversion from the potentiometer connected to it (see Fig. 4.4). Program 4.6 can read in any selected port to be read (PE0 to PE3). In the case, PE2 is being read at the address $1033 in the program. In program 4.7, we measure at pin6 of portA the sampling period.

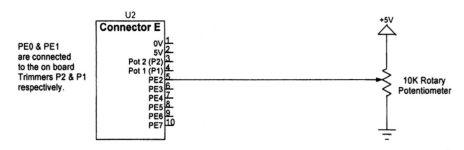

Fig. 4.4 A/D converter

Program 4.4: A/D Converter 1

```
            org  $0100
adctl   equ  $30         ; option registerr
addr    equ  $33          ; store conversion result
begin:  ldx  #$1000     ; reads in analog 0 – 5v into PE2 channel
            bset $39, x  $80
            bclr $39, x  $40
off:    ldab #$02
            stab adctl, x  ; select channel PE2 clr scan
check:  ldaa adctl, x  ; and mult bits of adctl
            brclr adctl, x $80 check   ; read status until conversion ends
            ldaa addr, x   ; If CCF=1 read results in ADR3 register
            staa $ec00
            bra  off       ; halt here
            org  $fffe
            fdb  begin
```

Program 4.5: A/D Converter 2

```
            org  $0100
begin:  ldx  #$1000   ; reads in analog 0 – 5v into PE2 channel
            bset $39, x  $80
            bclr $39, x  $40
off:    ldab #$02
            stab $1030   ; select channel PE2 clr scan
check:  ldaa $1030   ; and mult bits of adctl
            bpl  check   ; read status until conversion ends
            ldaa $1033   ; If CCF=1 read results in ADR3 register
            staa $ec00
            bra  off     ; halt here
            org  $fffe
            fdb  off
```

Program 4.6: A/D Converter 3

```
        org  $0100
adctl   equ  $30       ; option registerr
addr    equ  $33       ; store conversion result
begin:  ldx  #$1000    ; reads in analog 0 – 5v into PE2 channel
        bset $39, x  $80  ; bit in ADPU option reg to
        bclr $39, x  $40  ; power up and use E clk
off:    ldab #$30          ; use PE0-PE3 mult continuous scan=1
        stab adctl, x  ; select channel PE2 clr scan
again:  ldaa adctl, x  ; and mult bits of adctl
        brclr adctl, x  $80  again   ; read status until conversion ends
        ldaa addr, x   ;  channel PE2 at $1033
        staa $ec00
        bra  again     ; get next continuous scan
        org  $fffe
        fdb  begin
```

Program 4.7: A/D Converter 4

```
        org  $0100
adctl   equ  $30       ; This program measures
addr    equ  $33       ; the sampling period by
begin:  ldx  #$1000    ; sending 1 out pin A6
        ldaa #$40      ; pin 6
        bset $39, x  $80  ; bit in ADPU option reg to
        bclr $39, x  $40  ; power up and use E clk
off:    ldab #$30          ; use PE0-PE3 mult continuous scan=1
        stab adctl, x  ; select channel PE0 – PE3 clr scan
again:  ldaa adctl, x  ; and mult bits of adctl
        brclr adctl, x  $80  again    ; read status until conversion ends
        staa $1000     ; pulse A6
        eora #$40      ; reverse pulse A6
undone:bra  again      ; get next continuous scan
        org  $fffe
        fdb  begin
```

4.2   LCD and Keypad

Next, the LCD program 4.8 teaches how to program in assembly language the following extension .al1 assembly file. How to display a character of 8 bits word on LCD, via port M, 4 bits nibble at a time. You can replace "Welcome to the Lab" with own text. You also learn how to move the cursor to the start of each line. The flow chart illustrates the flow of steps involved in the program. The LCD used is a 16 × 2 characters module (Flowchart 4.1).

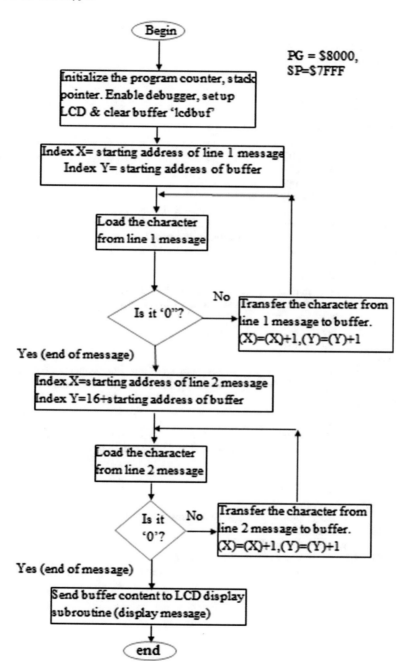

Flowchart 4.1 LCD display

Program 4.8: LCD Display

```
PORTB  equ  $1060
PORTC  equ  $1061
PORTM  equ  $1062
PORTN  equ  $1063
           sect  data
           org   $2000
           sect  text
           org   $8000
           lds   #$7FFF
           cli                        ; enable debugger
           bsr   lcdinit              ; initialize LCD
           ldx   #txt1                ; starting address of line 1 message --> index X
           ldy   #lcdbuf              ; starting address of buffer --> index Y
main1:
           ldab  0, x
           beq   main2                ; search for "0" – end of message
           stab  0, y        ; transfer line 1 into first 16 bytes of buffer
           inx
           iny
           bra   main1
main2:
           ldx   #txt2        ; starting address of line 2 message --> index X
           ldy   #lcdbuf+16   ; 16 + starting address of buffer --> index Y

main3:
           ldab  0, x
           beq   main4                ; search for "0" – end of message
           stab  0, y        ; transfer line 2 into next 16 bytes of buffer
           inx
           iny
           bra   main3
main4:
           bsr   lcdput       ; go subroutine for LCD display
end:       bra   end
          ; message for display
txt1:  fcb     'Welcome', 0 ; line 1 message
txt2:  fcb     'to the Lab.', 0      ; line 2 message
```

```
; **************Subroutine***********************
;  lcdnib,  lcdnibw – send code to LCD through PORTM
;  lcdtempo – delay
;  lcdputc – arrange command code to be sent
;  lcdinit – set up LCD
;  lcdput – arrange character code to be sent
             sect   data
lcdbuf:      rmb   32              ; reserve 32 bytes of ram as buffer for storing
                                   ; characters to be displayed
             sect   text

lcdnib:                           ; send code to LCD through PORTM (pm0 – pm3)
             andb   #$4F
             orab   #$30
             stab   PORTM
             tba
             oraa   #$80
             staa   PORTM         ; 1 to 500ns
             stab   PORTM
             rts
lcdnibw:
             bsr    lcdnib
             ldd   #20             ; delay 40us (20 x 2us)
             bsr    lcdtempo
             rts
lcdtempo:
             subd  #1             ; 1unit in reg D x 2us (12MHz)
             bne    lcdtempo
             rts
lcdputc:                          ; arrange command code – send 4 bits at one time
                                  ; to LCD screen, higher nibble goes first
             pshb                 ; retain the content of ACCB

             lsrb                 ; shift out lower nibble
             lsrb
             lsrb
             lsrb
             bsr    lcdnib; send higher nibble
             pulb                 ; restore ACCB
             bsr    lcdnibw       ; send lower nibble
             rts
lcdinit:                          ; set up the LCD for 4 lines interface, 2 line mode,
                                  ; 5x7 dot format, increment, display shift off.
             Ldab  #$3F
             stab   PORTM
             ldd   #8000
             bsr    lcdtempo      ; 15ms
             ldab  #3
```

```
            bsr   lcdnib          ; LCD <-- 3
            ldd   #2000
            bsr   lcdtempo        ; 4ms
            ldab  #3
            bsr   lcdnib          ; LCD <-- 3
            ldd   #50
            bsr   lcdtempo        ; 0.1ms
            ldab  #3
            bsr   lcdnibw         ; LCD <-- 3, 40us
            ldab  #2
            bsr   lcdnibw         ; LCD <-- 2, 40us
            ldab  #$29
            bsr   lcdputc         ; LCD <-- 29
            ldab  #$08
            bsr   lcdputc         ; LCD <-- 08, display off
            ldab  #$01
            bsr   lcdputc         ; LCD <-- 01, clear
            ldd   #2000
            bsr   lcdtempo
            ldab  #$06
            bsr   lcdputc         ; LCD <-- 06
            ldab  #$0E
            bsr   lcdputc         ; LCD <-- 0E, display on, cursor
            ldx   #lcdbuf               ; clear the buffer
            ldab  #' '
lcdin1:     stab  0, x
            inx
            cpx   #lcdbuf+32
            bne   lcdin1
            rts
lcdput:                           ; transfer content of buffer to LCD
            ldab  #$80            ; move cursor to the beginning of first line
            bsr   lcdputc
            ldx   #lcdbuf
lcdp1:                            ; cursor at end of line 1?
            Cpx   #lcdbuf+16
            bne   lcdp2
            ldab  #$A8            ; if yes, move cursor to the start of 2nd line
            bsr   lcdputc
lcdp2:      ldab  0, x            ; arrange and send the character code – 4 bits at one
                                  ; time, higher nibble goes first
```

```
lsrb
lsrb
lsrb
lsrb
orab   #$40
bsr    lcdnib ; send higher nibble
ldab   0, x
orab   #$40
bsr    lcdnibw        ; send lower nibble
inx
cpx    #lcdbuf+12
bne    lcdp1
rts
```

The next program for the keypad and LCD demonstrates how to read in characters from Port D & G of connector D, 4 bits at a time. You will learn how to
change the content of the conversion table and check the LCD character map. The
keypad used is a 16 keys keypad. We use a keypad encoder as shown in Fig. 4.5,
for the keypad connections (Flowchart 4.2).

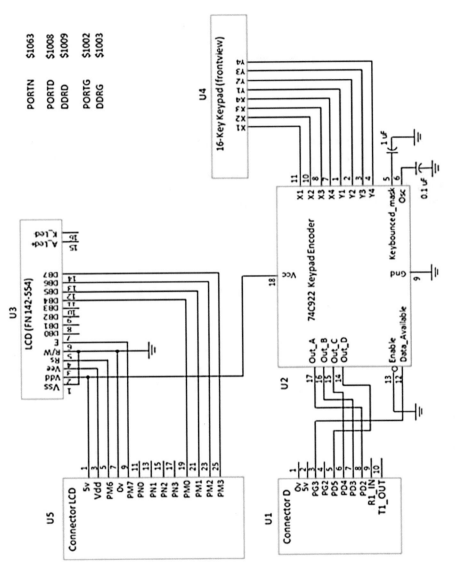

Fig. 4.5 Keypad and LCD using keypad encoder

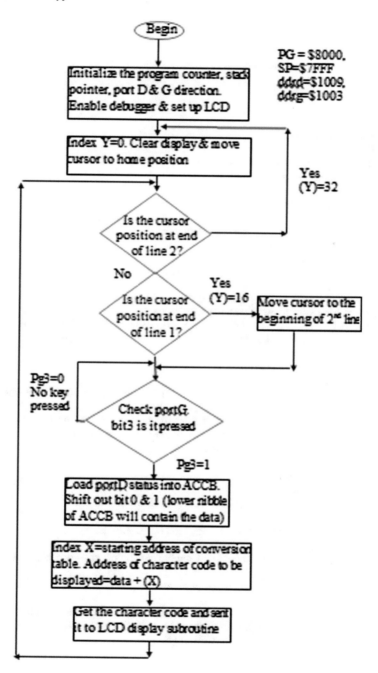

Flowchart 4.2 Keypad and LCD

Program 4.9: Keypad & LCD

```
PORTD  equ   $1008
DDRD   equ   $1009
PORTG  equ   $1002
DDRG   equ   $1003
PORTM  equ   $1062
; initialize the program counter, stack pointer, LCD,
; Port D and Port G direction registers.
         sect   text
         org    $8000        ; program starts at $8000
         lds    #$7FFF
         cli                  ; enable debugger
         bsr    lcdinit  ; initialize LCD
         ldab   #$00
         stab   DDRD         ; set portd as inputs
         ldab   #$03
         stab   DDRG         ; set portg bit 2–7 as inputs
; start checking keypad entry and display characters
start:   ldy    #0           ; use index Y to keep track the cursor position
         ldab   #$01  ; clear display & home cursor: command code "$01"
         bsr    lcdputc
         ldd    #2000  ; delay
         bsr    lcdtempo
loop:    cpy    #32          ; checking cursor position – is it at the end of line 2?
         beq    start        ; if "Yes", clear and home
         cpy    #16          ; checking cursor position – is it at the end of line 1?
         beq    loop1  ; if "Yes", move cursor to the beginning of line 2 –
                             ; command code "$A8"
         bra    next
loop1:   ldab   #$A8
         bsr    lcdputc
next:    ldaa   PORTG        ; check data available signal at pg3
         staa   $50
         brclr  $50   $08   next   ; "high" to ensure valid keypad entry
         ldab   PORTD        ; read portd
         andb   #%00111100        ; keep the content of pd2 – pd5 and mask out
                                   ; the rest
         lsrb ; shift pd2 -- pd5 to lower nibble
         lsrb
         ldx    #table1      ; starting address of conversion table -->index X
         abx                 ; get the actual character to be displayed, (X) +
                             ; (ACCB) --> (X)
```

```
            ldab  0, x  ; load character code from conversion table and
            bsr   lcdput_keypad      ; branch to subroutine for LCD display
            ldd   #$ffff ; delay – prevent multiple entry
            bsr   lcdtempo
            ldd   #$ffff ; delay again
            bsr   lcdtempo
            iny
            bra   loop
; set the conversion table
table1:  fcb  $31, $32, $33, $46, $34, $35, $36, $45, $37, $38, $39, $44, $41,
             $30, $42, $43

;**************Subroutine***************************
; lcdnib, lcdnibw – send code to LCD through PORTM
; lcdtempo – delay
; lcdputc – arrange command code to be sent
; lcdinit – set up LCD
; lcdput_keypad – arrange character code to be sent
lcdnib:                 ; send code to LCD through PORTM (pm0 – pm3)
            andb  #$4F
            orab  #$30
            stab  PORTM
            tba
            oraa  #$80
            staa  PORTM
            stab  PORTM
            rts
lcdnibw:
            bsr   lcdnib
            ldd   #20
            bsr   lcdtempo
            rts
lcdtempo:                   ; delay
            subd  #1
            bne   lcdtempo
            rts
lcdputc:                ; arrange command code – send 4 bits at one time,
                        ; higher nibble first
            pshb; retain the content of ACCB
            lsrb
            lsrb
            lsrb
            lsrb
            bsr   lcdnib ; send higher nibble
            pulb           ; restore ACCB
            bsr   lcdnibw          ; send the lower nibble
            rts
```

```
lcdinit:                        ; set up the LCD for 4 lines interface, 2 line mode,
                                ; 5x7 dot format, increment, display shift off
        ldab #$3F
        stab  PORTM
        ldd   #8000
        bsr   lcdtempo
        ldab #3
        bsr   lcdnib
        ldd   #2000
        bsr   lcdtempo
        ldab #3
        bsr   lcdnib
        ldd   #50
        bsr   lcdtempo
        ldab #3
        bsr   lcdnibw
        ldab #2
        bsr   lcdnibw
        ldab #$29
        bsr   lcdputc
        ldab #$08
        bsr   lcdputc
        ldab #$01
        bsr   lcdputc
        ldd   #2000
        bsr   lcdtempo
        ldab #$06
        bsr   lcdputc
        ldab #$0E
        bsr   lcdputc
        rts
lcdput_keypad:                  ; arrange character code – send 4 bits at a time, higher
                                ; nibble first
        stab   $0051 ; keep content of ACCB in location $0051
        lsrb                    ; shift out lower nibble
        lsrb
        lsrb
        lsrb
        orab   #$40
        bsr    lcdnib  ; send higher nibble
        ldab   $0051; restore ACCB
        orab   #$40
        bsr    lcdnibw            ; send lower nibble
        rts
```

The following program (4.10) in C language performs the function of calculating the square root of a number we key in. You may read the input from the keypad, and the answer for the square root will display on the LCD screen. As shown in the diagram, the 16-key keypad input goes into the portN and portM of the micro-controller connector. The connector LCD links to portM and portN of the micro-controller. The board trimmer P3 controls the brightness of the LCD. The answer in float number will display via portM, 4 bit (nibble) at a time. Compare the schematic

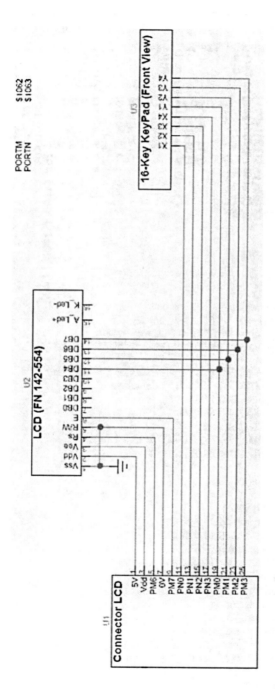

Fig. 4.6 Keypad and LCD

diagram with the other to find out the complexities of the other keypad and LCD circuit diagram. You may also note the simplicity of using the high levels language instead of the assembly language (Fig. 4.6).

```
------------------------------------------------------------
Program 4.10: Keypad & LCD In C Program
------------------------------------------------------------
#include "hc11.h"
#define  PORTM  *(unsigned char *)(_IO_BASE + 0x62)
#define  PORTN  *(unsigned char *)(_IO_BASE + 0x63)
#include <stdio.h>
#include <math.h>
void  lcdinit(void);
char  keyget(void);
double  e;

void main(void)
{    int  i;
     lcdinit( );
     for (i=2;;)
{        e = sqrt(i);
         if (i>=0)
          print(" sqrt(%d)=%f", i, e);
          i= keyget( ) – '0';   }  }

/************************Lcd***************************
//PORTM bit7=LCD EN = 0 by default
//PORTM bit4=DAC CS = 1 by default
#define  LCDCOLS  16
#define  LCDLINES  2
#define  LCDBUFSZ (LCDCOLS*LCDLINES)
char  lcdbuf[LCDBUFSZ];

void  lcdwait(int cnt)
{   int  i;
    for (; cnt; cnt--)
    for (i=0; i<1000; i++);    }
void  lcdnib( unsigned char c)
{        c = (c & 0x4F) | 0x30;
         PORTM = c | 0x80;  /*1 x 500ns*/
         PORTM = c;    }
void  lcdnibw(unsigned char c)
{        lcdnib (c);
         lcdwait(0);    }
void  lcdputc(unsigned char c)
{        lcdnib((c>>4)&0xF);
          lcdnibw(c&0xF);        }
void  lcdinit(void)
{        char  *s;
         PORTM = 0x3F;
         lcdwait(5);    /* 15ms */
```

```
          lcdnib(3);        /* lcd = 3 */
          lcdwait(1);       /* 4.1ms */
          lcdnib(3);        /* lcd = 3 */
          lcdwait(0);       /* 0.1 ms */
          lcdnibw(3);       /* lcd = 3 x 40us */
          lcdnibw(2);       /* lcd = 2 x 40us */
          lcdputc(0x28 | (LCDLINES-1)); /*LCDLINES:info for the sim11*/
          lcdputc(0x08);           /* display off */
          lcdputc(0x01);           /* clear display */
          lcdwait(1);
          lcdputc(0x06);
          lcdputc(0x0E);           /* display on, cursor */
          for (s=lcdbuf; s<lcdbuf + LCDBUFSZ; s++)
          *s = ' ';
          putchar(' ');     }
int  putchar(char c)    /*library function*/
{      char *s;
       for (s=lcdbuf; s<lcdbuf + LCDBUFSZ – 1; s++)
               *s = s[1];
               *s = c;
       lcdputc(0x80);
       for (s = lcdbuf; s<lcdbuf + LCDBUFSZ; s++)
       {      if (s==lcdbuf + (LCDBUFSZ/2))
              lcdputc(0x80 + 0x28);
              lcdnib( (*s >>4) | 0x40);
              lcdnibw(*s | 0x40);    }
       return c;   }


/*************Keyboard***********************/

char  keyget(void)
{      static char old_key;
       char  x = 0;
#if 0  // keypad 3 x4
       PORTM = 0x37;                    /* K=0 J=1 H=1 G=1 */
       if ((PORTN&0x08)==0) x = '1';
       if ((PORTN&0x04)==0) x = '2';
       if ((PORTN&0x02)==0) x = '3';
       PORTM = 0x3B;                    /* K=1 J=0 H=1 G=1 */
       if ((PORTN&0x08)==0) x = '4';
       if ((PORTN&0x04)==0) x = '5';
       if ((PORTN&0x02)==0) x = '6';
       PORTM = 0x3D;                    /* K=1 J=1 H=0 G=1 */
       if ((PORTN&0x08)==0) x = '7';
       if ((PORTN&0x04)==0) x = '8';
       if ((PORTN&0x02)==0) x = '9';
       PORTM = 0x3E;                    /* K=1 J=1 H=1 G=0 */
```

```
        if ((PORTN&0x08)==0) x = '*';
        if ((PORTN&0x04)==0) x = '0';
        if ((PORTN&0x02)==0) x = '#';
#else  // keypad 4 x 4
        PORTM = 0x37;              /* K=0 J=1 H=1 G=1 */
        if ((PORTN&0x08)==0) x = 'C';
        if ((PORTN&0x04)==0) x = 'B';
        if ((PORTN&0x02)==0) x = '0';
        if ((PORTN&0x01)==0) x = 'A';
        PORTM = 0x3B;              /* K=1 J=0 H=1 G=1 */
        if ((PORTN&0x08)==0) x = 'D';
        if ((PORTN&0x04)==0) x = '9';
        if ((PORTN&0x02)==0) x = '8';
        if ((PORTN&0x01)==0) x = '7';
        PORTM = 0x3D;              /* K=1 J=1 H=0 G=1 */
        if ((PORTN&0x08)==0) x = 'E';
        if ((PORTN&0x04)==0) x = '6';
        if ((PORTN&0x02)==0) x = '5';
        if ((PORTN&0x01)==0) x = '4';
        PORTM = 0x3E;              /* K=1 J=1 H=1 G=0 */
        if ((PORTN&0x08)==0) x = 'F';
        if ((PORTN&0x04)==0) x = '3';
        if ((PORTN&0x02)==0) x = '2';
        if ((PORTN&0x01)==0) x = '1';
#endif
        if  (x == old_key)
        return  0;
        old_key  = x;  return x;                        }
```

4.3 Waveform Timings

We can appreciate the use of the microcontroller (68HC11) timing system to generate a square waveform. The program 4.11 shows the square wave generator. The square wave generated from portA bit 5 (PA5) can display on the oscilloscope. We alter the period of the waveform by changing the value '#4000' in the interrupt subroutine.

Program 4.11: Square Wave Timer

```
regbas  equ  $1000   ; bas address
porta   equ  $0      ; port A
tcnt    equ  $0e     ; timer counter
toc3    equ  $1a     ; timer output compare 3
tmsk1   equ  $22     ; timer interrupt mask 1
tflg1   equ  $23     ; timer interrupt flag 1
tctl1   equ  $20     ; timer control 1
; outputsquare wave signal at bit5 of porta
        org  $8000   ; program starts at $8000
        cli
main:   ldx  #regbas
        ldaa #%00100000
        staa tmsk1, x ; OC3I interrupt enable
        staa tflg1, x ; send a one to clear the OC3F flag
        ldaa #%00010000
        staa tctl1, x ; OM3:OL3=0:1;toggle TOC3 output after a successful compare
        ldd  tcnt, x  ; load current content of timer counter
        addd #2000    ; added by 2000 counts (333nsec per count, total up 0.67msec)
        std  toc3, x  ; store the result in compare register TOC3
undone: bra  undone   ; do something useful here

;***********interrupt subroutine*****************************
rtoc:   bset tflg1, x $20   ; send a "one" to clear the OC3F flag
        ldd  #4000    ; set high or low period for 400x333 = 1.332msec
        addd tcnt, x  ; store the value in compare register
        std  toc3, x  ; interrupr generated when next tcnt = toc3
        rti
;*************define interrupt vector address*****************
org  $ffe4
fdb  rtoc
```

We can program a pulse width modulation signal to drive a DC motor. The output for portA bit5 of the microcontroller connects to the motor. Observe that we adjust the speed of the motor by varying "hi_time" between 1000 and 7000. At the label low in the interrupt subroutine, the value #7500-hi_time is added to Accumulator D.

Program 4.12: PWM Signal of 400Hz at 2.5ms Period

```
regbas  equ  $1000   ; bas address
porta   equ  $0      ; port A
tcnt    equ  $0e     ; timer counter
toc3    equ  $1a     ; timer output compare 3
tmsk1   equ  $22     ; timer interrupt mask 1
tflg1   equ  $23     ; timer interrupt flag 1
tctl1   equ  $20     ; timer control 1
hi_time equ  3000    ; high period 3000 cnts (333ns per count)
        org  $8000   ; program starts at $8000
        cli
main:   bsr   initoc3 ; set up timer output compare 3, TOC3
undone: nop
        bra  undone

;******************subroutine*********************
Initoc3:      ldx   #regbase
        ldaa  #$20
        staa  porta, x  ; set PA5 high first
        staa  tmsk1, x  ; OC3I interrupt enable
        staa  tflg1, x  ; send a "one" to clear OC3F interrupt flag
        staa  tctl1, x  ; OM3: OL3 = 1:0; set PA5 low upon interrupt
        ldd   tcnt, x   ; load current content of timer counter
        addd  #50       ; added by 50 counts, initial setup before interrupt occurs
        std   toc3, x   ; store the result in compare register TOC3
        rts

;***********interrupt subroutine*****************************
rtoc3:  ldx   #$1000
        ldaa  porta, x  ; check porta PA5 high or low
        bita  #$20
        beq   low
high:   ldd   tcnt, x
```

```
        addd  #hi_time; set the new value to be compared = high period
        std   toc3, x  ; store the value in TOC3 register
        ldaa  #$20
        staa  tcntl1, x ; OM3 : OL3 =1:0; set PA5 low upon interrupt
        bra   undone
  low:  ldd   tcnt, x          ; period of 2.5ms (7500 x 333ns)
        addd  #7500-hi_time  ; set the new value to be compared = low period
        std   toc3, x  ; store the value in TOC3 register
        ldaa  #$30    ;
        staa  tcntl1, x ; OM3 : OL3 =1:1; set PA5 high upon next OC3 interrupt
  done: ldaa  #$20     ; clear OC3F interrupt flag
        staa  tflg1, x
        rti

;**************define interrupt vector address*******************
org   $ffe4
fdb   rtoc3
```

The next program can perform three functions with the external connections. Refer to the circuit diagram for the program. The optical or slotted switch (connected to PA0) is used as an input device to start the motor. After starting the motor, the microcontroller can ignore the input signal from the slotted switch. Note 1: It is important to note that to establish the correct initial status of the slotted switch the power supply to the optical switch and motor must be turned on before turning on the microcontroller. Note 2: After you have loaded the program, sometimes pressing 'Go' does not work as expected. If this happens, press the Reset switch instead of 'Go'.

The external rotary potentiometer connects to the A/D converter (PE2) port E. We use the potentiometer to control the motor speed. Depending on the converted 8 bit reading at PE2, three distinct speed levels are set accordingly.

PE2 reading (Hex)	Value of "h_time"
00–40	1000
41–C0	4000
C1–FF	7000

Note: Under the label 'low' in the interrupt subroutine, the value (#7500-hi_time) should be added to Accumulator D. You may refer to the diagram in Fig. 4.7 for the PWM control of DC motor. The A/D pulse width varying circuit is the same as Fig. 4.4 A/D converter. Port E is only a 8 bit read in port for A/D converter. The third function is the inclusion of the XIRQ interrupt subroutine to stop the motor. We connect a push button to the X connector's XIRQ pin of the microcontroller unit.

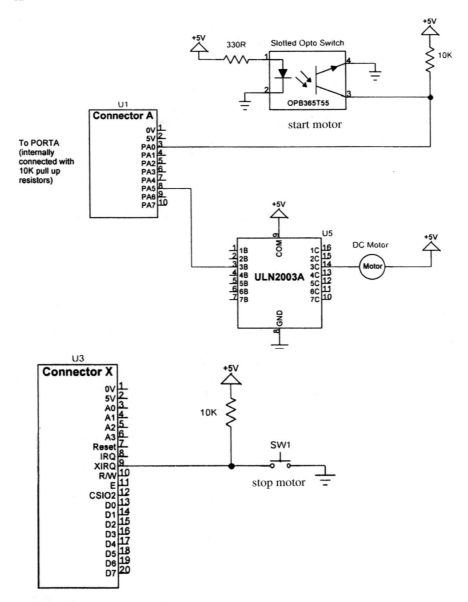

Fig. 4.7 PWM control of DC motor

Program 4.13: PWM Motor Control With A Potentiometer

```
; pwm output at porta bit5 (PA5)
; pulse period of 2.5ms (7500 cnts x 333ns), pulse frequency is 400Hz
regbas  equ  $1000   ; base address
porta   equ  $0      ; port A
ddra    equ  $1
tcnt    equ  $0e     ; timer counter
toc3    equ  $1a     ; timer interrupt mask 1
tflg1   equ  $23     ; timer control 1
adct1   equ  $30     ; ADCTL register
adr4    equ  $34     ; A?D result register
optn    equ  $39     ; option register
portg   equ  $02
ddrg    equ  $03
hi_time equ  3000    ; high period, 3000 cnts (333ns per count)
org     $8000  ; program starts at $8000
clra
tap     ; transfer ACCA to CCR – to clear interrupt M
cli
begin:  ldx    #regbase
        ldaa   #$FF
        staa   ddrg, x
        ldab   #$FF
        stab   portg, x  ; turn off red and green LEDs
        bset   optn,x $80; enable A/D conversion system – set bit 7
        bsr    delay    ; delay of 100 ms is required to stabilize A/D converter
        bclr   optn, x $40 ; use sys clock E
        ldab   #$22     ; perform continuous scan & single channel conversion (PE2
only)
        stab   adctl,x  ; at the same time, start the conversion by writing to ASCTL reg.
        ldaa   #$20
        staa   tctl1, x
swtich1: brclr porta, x $01 switch1    ; check opto switch
switch2: brset porta, x $01 switch2    ; off-on-off sequence
main:   bsr   initoc3 ; set up timer output compare 3, TOC3
undone: nop   ; do something interesting here
        bra    undone
;********************subroutine*****************
Initoc3: ldx    #regbase
        ldaa   #$20
        staa   porta, x  ; set PA5 high first
        staa   tmsk1, x ; OC3I interrupt enable
        staa   tflg1, x  ; send a "one" to clear OC3F interrupt flag
        staa   tctl1, x  ; OM3:OL3=1:0; set PA5 low upon interrupt
```

```
           ldd    tcnt, x    ; load current content of timer counter
           addd   #50        ; added by 50 counts, initial setup before interrupt occurs
           std    toc3, x    ; stored result in TOC3 compare reg.
           rts
delay:     ldaa   #240
loop:      suba   #1
           bne    loop
           rts
;************interrupt subroutine******************
rtoc3:     ldx    #regbas
again:     brclr  adctl, x $80 again    ; check status and wait for end of conversion – bit7
           ldaa   adr4, x   ; transfer the converted value from A/D result register ADR4
           cmpa   #$C0      ; check PE2 >= 3.75v
           bhs    sp3
           cmpa   #$80      ; check PE2 >=2.5v
           bhs    sp2
           cmpa   #$40      ; check PE2 >= 1.25v
           bhs    sp1
sp1:       ldd    #1000 ; speed 1 'ON' width
           std    $0050
           ldd    #6500 ; speed 1 'OFF' width
           std    $0052
           bset   portg, x  $03
           bra    next
sp2:       ldd    #4000 ; speed 2 'ON' width
           std    $0050
           ldd    #3500 ; speed 2 'OFF' width
           std    $0052
           bset   portg, x  $03
           bclr   portg, x  $01
           bra    next
sp3:       ldd    #7000 ; speed 3 'ON' width
           std    $0050
           ldd    #500   ; speed 3 'OFF' width
           std    $0052
           bclr   portg, x  $03
next:      ldaa   porta, x; check PA5 high or low
           bita   #$20
           beq    low
high:      ldd    tcnt, x
           addd   $0050 ; set the new value to be compared = high period
           std    toc3, x; store the value in TOC3 register
           ldaa   #$20
           staa   tctl1, x ; OM3:OL3=1:0; set PA5 low upon interrupt
           bra    done
low:       ldd    tcnt, x
           addd   $0052 ; set the new value to be compared = low period
```

```
        std     toc3, x ; store the value in TOC3 register
        ldaa    #$30
        staa    tctl1, x ; OM3:OL3=1:1;next OC3 interrupt set pin PA5 high
done:   ldaa    #$20    ; clear OC3F interrupt flag
        staa    tflg1, x
        rti
;***********interrupt subroutine (XIRQ)******************
Xirq_sub:       ; do something here
        Ldx     #regbas
        Ldaa    #$20    ; PA5 low to stop motor
        staa    tctl1, x
        bclr    portg, x $03
check1: brclr   porta, x $01   check1 ; wait for opto to sense then start motor again
check2: brset   porta, x $01   check2 ; off-on-off sequence
        bset    portg, x $03
        rti
;*******define interrupt vector address***************
        org     $ffe4
        fdb     rtoc3
        org     $fff4   ; XIRQ (active low)
        fdb     xirq_sub
```

4.4   Pressure Sensing

Smart sensors can adjust their interface outputs to the receiving appliances. But there are factors exist that produce errors in the measuring sensors. Factors to consider are scaling (static range) and offset voltages, drift due to temperature after prolonged operations in a non-cooled space, noise margins and finally, bandwidth for dynamics, frequency-related measurements. Therefore, signal conditioning circuits are often necessary for better interpretation and presentation of the incoming signal from the sensors into the microcontroller.

The setup in the figure measures the air pressure in the host line using the SCX01DNC pressure transducer. Verify the resistors are correct to give the desired output equations. We can vary the input pressure and plot the digital values

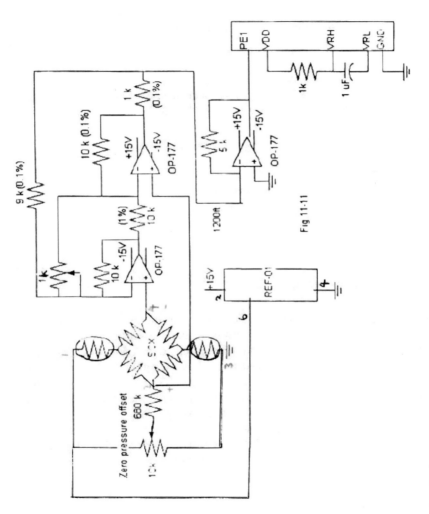

Fig. 4.8 Pressure sensing circuit

converted by the 68HC11 A/D converter. The oscilloscope can be used to find the time delay from the instance of pressure change to the time of complete digital conversion by the processor, that is, the duty cycle of the system. Most A/D converters lose their accuracy when the voltages come close to the rail values. We can improve it by changing the resolution to the A/D converter. The circuit for the pressure detector requires a gain to amplify the signal large enough to view or use for other activities. The amplifiers simply do the job. Any of the A/D converter programs (P4.4–P4.7) can be used to measure the converted signal values (Fig. 4.8).

4.5 Temperature Measurement

Why is the standard 741 Op Amp not good enough? It gives an offset of 6 mV and drift of 15 mV/°C. The BJT is also susceptible to noise from RF rectification. So, let say we want to amplify the signal to a magnitude of 100, for a temperature sensing of 100 mV/°C. We will have an offset of 6 °C and a drift of ±7 °C! We must use a more precise CMOS-based amplifier. The circuit for the temperature measurement is as shown in the diagram. AD590 is the temperature sensor. The resistors are chosen to give a slope of 5. Trim pots are necessary to ensure this as well as to remove offset voltage at 0 °C. Use a hot blower to increase the measured temperature. We measure the Vo into the portE of the microcontroller. $Vo = (50 \text{ mV/°C}) * T$, where T represents the temperature measured. We can make use of any of the A/D converter program 4.4–4.7 to detect the signal into the microcontroller unit. We can then compare and verify the voltage signal with a thermometer (Fig. 4.9).

4.6 Stepper Motor Control

We can program the 68HC11 microcontroller unit to drive a stepper motor. The stepper motor we use needs a 5 V input power supply. Power MOSFETs for example, the BUZ12 n-channel transistors are used to switch on the stepper motor. The terminals of Port A of the MCU connects to output to the gates of each transistor. The stepper motor has four coils to be activated in a sequential

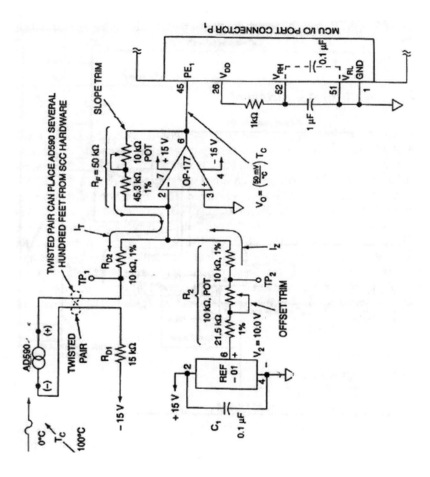

Fig. 4.9 Temperature sensing circuit

manner to rotate the motor. IN4001 diodes are used as free-wheeling diodes to protect the motor coils when the transistors are de-activated. It helps to relieve the stored charges in the coils when the transistors are turned off. So the motor coils are protected from burnings. Three methods used can activate and energise the motor. They are the full phase, full step and half step modes. The three programs written in assembly language for the 68HC11 microcontroller illustrates the different techniques used to energise the four motor coils for rotating the motor. Only the correct sequence of turning on the phase of the coils will activate the motor to turn in a continuous direction. The 68HC12 family has slightly different configurations in the registers. We can just change the data direction register, and the data port register addresses in the program, to perform the same function (Fig. 4.10).

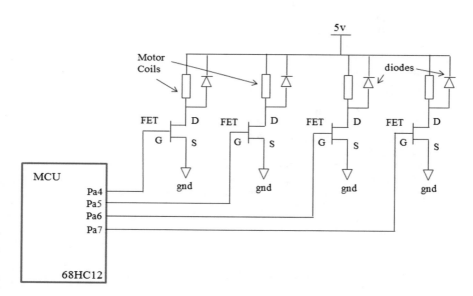

Fig. 4.10 Stepper motor driving circuit

Program 4.14: Two Phase Mode

```
;Program to drive a 200step/rev unipolar stepper motor in two-phase on
;mode(200steps/rev).
;Two-phase on mode gives higher torque at the expense of higher current.
;two-phase on mode(cw) = coilsA+B then coilsB+C then coilsC+D then
;coilsD+A then repeat cycle.
;two-phase on mode(ccw)= coilsC+D then coilsB+C then coilsA+B then
;coilsD+A then repeat cycle.
;The motor shaft rotates clockwise by 1rev(200steps) then ;counterclockwise by
;1rev
;A short delay separates each cw and ccw motion.
;This is repeated for ever.
;The stepping rate (steps/sec) is set by the delay subroutine
;coilA connected to pa7
;coilB connected to pa6
;coilC connected to pa5
;coilD connected to pa4
;start initialisation
        org     $e000
        cli
; set bit7 of PACTL to output
        ldaa    #$80
        staa    $1026 ;bit7=1 in DDRA7
;finish initialisation
;rotate shaft cw by 1 rev (200 steps)
;use the Y index reg as a counter
again   ldy     #50     ;200steps=50 x 4 where 50 is loaded into IY and
                        ;4=four steps of cycle
revcw   jsr     cw
        dey
        bne     revcw


;now give short delay before reversing (share the use of the delay subroutine)
        ldy     #25
shdel1  jsr     delay
        dey
        bne     shdel1
;now rotate shaft ccw by 1 rev (200 steps)
;again use the Y index reg as a counter
        ldy     #50
revccw          jsr     ccw
        dey
        bne     revccw
```

```
;now give another short delay before reversing again(share the use of the delay
;subroutine)
        ldy    #25
shdel2 jsr    delay
        dey
        bne    shdel2
        jmp    again
;subroutine to step through 1 cycle of full-step mode in
;clockwise direction
cw      ldaa   #$c0   ;step1 of cycle(coilsA+B=on)
        staa   $1000
        jsr    delay
        ldaa   #$60   ;step2 of cycle(coilsB+C=on)
        staa   $1000
        jsr    delay
        ldaa   #$30   ;step3 of cycle(coilsC+D=on)
        staa   $1000
        jsr    delay
        ldaa   #$90   ;step4 of cycle(coilsD+A=on)
        staa   $1000
        jsr    delay
        rts
;subroutine to step through 1 cycle of full-step mode in
;counter clockwise direction
ccw     ldaa   #$30   ;step3 of cycle(coilsC+D=on)
        staa   $1000
        jsr    delay
        ldaa   #$60   ;step2 of cycle(coilsB+C=on)
        staa   $1000
        jsr    delay
        ldaa   #$c0   ;step1 of cycle(coilsA+B=on)
        staa   $1000
        jsr    delay
        ldaa   #$90   ;step4 of cycle(coilsD+A=on)
        staa   $1000
        jsr    delay
        rts
;delay routine works close to pull-in speed of motor
delay  ldx    #500   ;
del1   dex
        bne    del1
        rts
```

Program 4.15: Full Step Mode

;Program to drive a 200step/rev unipolar stepper motor in full-step mode.
;Full-step mode(cw) = coilA then coilB then coilC then coilD then repeat cycle.
;Full-step mode(ccw)= coilD then coilC then coilB then coilA then repeat cycle
;The motor shaft rotates clockwise by 1rev(200steps) then counterclockwise by
;1rev
;A short delay separates each cw and ccw motion
;This is repeated for ever.
;The stepping rate (steps/sec) is set by the delay subroutine
;coilA connected to pa7
;coilB connected to pa6
;coilC connected to pa5
;coilD connected to pa4
;start initialisation
```
        org     $e000
        cli
; set bit7 of PACTL to output
        ldaa    #$80
        staa    $1026 ;bit7=1 in DDRA7
;finish initialisation
;rotate shaft cw by 1 rev (200 steps)
;use the Y index reg as a counter
again  ldy     #50     ;200steps=50 x 4 where 50 is loaded into IY and
                       ;4=four steps of cycle
revcw  jsr     cw
       dey
       bne     revcw
;now give short delay before reversing (share the use of the delay subroutine)
       ldy     #25
sdel1  jsr     delay
       dey
       bne     sdel1
;now rotate shaft ccw by 1 rev (200 steps), again use the Y index reg as a counter
       ldy     #50
evccw  jsr     ccw
       dey
       bne     evccw
```
;now give another short delay before reversing again(share the use of the delay
subroutine)

```
        ldy     #25
sdel2   jsr     delay
        dey
        bne     sdel2
        jmp     again
;subroutine to step through 1 cycle of full-step mode in
;clockwise direction
cw      ldaa    #$80    ;step1 of cycle(coilA=on)
        staa    $1000
        jsr     delay

        ldaa    #$40    ;step2 of cycle(coilB=on)
        staa    $1000
        jsr     delay
        ldaa    #$20    ;step3 of cycle(coilC=on)
        staa    $1000
        jsr     delay
        ldaa    #$10    ;step4 of cycle(coilD=on)
        staa    $1000
        jsr     delay
        rts
;subroutine to step through 1 cycle of full-step mode in
;counter clockwise direction
ccw     ldaa    #$20    ;step3 of cycle(coilC=on)
        staa    $1000
        jsr     delay
        ldaa    #$40    ;step2 of cycle(coilB=on)
        staa    $1000
        jsr     delay
        ldaa    #$80    ;step1 of cycle(coilA=on)
        staa    $1000
        jsr     delay
        ldaa    #$10    ;step4 of cycle(coilD=on)
        staa    $1000
        jsr     delay
        rts
;delay routine works close to pull-in speed of motor
delay   ldx     #500    ;
del1    dex
        bne     del1
        rts
```

Program 4.16: Half Step Mode

```
;Program to drive a 200step/rev unipolar stepper motor in half-step mode
(400steps/rev).
;The motor runs smoother in half-step mode.
;Half-step mode(cw) = coilA then coilsA+B then coilB then coilsB+C
;then coilC then coilsC+D then coilD then coilsD+A then repeat cycle.
;Half-step mode(ccw)= coilD then coilsC+D then coilC then coilsB+C
;then coilB then coilsA+B then coilA then coilsD+A thenrepeat cycle.
;The motor shaft rotates clockwise by 1rev(400steps) then counterclockwise by
;1rev(another 400steps)
;A short delay separates each cw and ccw motion
;This is repeated for ever.
;The stepping rate (steps/sec) is set by the delay subroutine
;coilA connected to pa7
;coilB connected to pa6
;coilC connected to pa5
;coilD connected to pa4
;start initialisation
        org    $e000
        cli
;set bit7 of PACTL to output
        ldaa   #$80
        staa   $1026 ;bit7=1 in DDRA7
;finished initialisation
;now rotate shaft cw by 1 rev (400 steps)
;use the Y index reg as a counter
again   ldy    #50    ;400steps=50 x 8 where 50 is loaded into IY and
                      ;8=eight steps of cycle
rvcw    jsr    cw
        dey
        bne    rvcw
;now give short delay before reversing (share the use of the delay subroutine)
        ldy    #100
hdel1   jsr    delay
        dey
        bne    hdel1
;now rotate shaft ccw by 1 rev (400 steps)
;again use the Y index reg as a counter
        ldy    #50
revccw  jsr    ccw
        dey
        bne    revccw
;now give another short delay before reversing again(share the use of the delay
subroutine)
```

```
          ldy    #100
hdel2  jsr    delay
          dey
          bne    hdel2
          jmp    again
;subroutine to step through 1 cycle of half-step mode in
;clockwise direction
cw     ldaa   #$80   ;step1 of cycle(coilA=on)
          staa   $1000
          jsr    delay
          ldaa   #$c0   ;step2 of cycle(coilA+B=on)
          staa   $1000
          jsr    delay
          ldaa   #$40   ;step3 of cycle(coilB=on)
          staa   $1000
          jsr    delay
          ldaa   #$60   ;step4 of cycle(coilB+C=on)
          staa   $1000
          jsr    delay
          ldaa   #$20   ;step5 of cycle(coilC=on)
          staa   $1000
          jsr    delay
          ldaa   #$30   ;step6 of cycle(coilC+D=on)
          staa   $1000
          jsr    delay
          ldaa   #$10   ;step7 of cycle(coilD=on)
          staa   $1000
          jsr    delay
          ldaa   #$90   ;step8 of cycle(coilD+A=on)
          staa   $1000
          jsr    delay
          rts
;subroutine to step through 1 cycle of full-step mode in
;counter clockwise direction
ccw    ldaa   #$10   ;step7 of cycle(coilD=on)
          staa   $1000
          jsr    delay
          ldaa   #$30   ;step6 of cycle(coilC+D=on)
          staa   $1000
          jsr    delay
          ldaa   #$20   ;step5 of cycle(coilC=on)
          staa   $1000
          jsr    delay
          ldaa   #$60   ;step4 of cycle(coilB+C=on)
          staa   $1000
          jsr    delay
          ldaa   #$40   ;step3 of cycle(coilB=on)
```

```
         staa    $1000
         jsr     delay
         ldaa    #$c0   ;step2 of cycle(coilA+B=on)
         staa    $1000
         jsr     delay
         ldaa    #$80   ;step1 of cycle(coilA=on)
         staa    $1000
         jsr     delay
         ldaa    #$90   ;step8 of cycle(coilD+A=on)
         staa    $1000
         jsr     delay
         rts
;delay routine works close to pull-in speed of motor note that pull-in speed of
;half step mode is approx twice that of full-step mode but note that angular
;velocity of pull-in speed for both modes is approx the same.
delay    ldx     #250  ;
del1     dex
         bne     del1
         rts
```

4.7 Serial Communications

We can use a computer to communicate with the microprocessor via RS232 connect
to COM1 of the microprocessor unit. A simple experiment starts by opening the
hyper terminal for the interface. We load a program into the microprocessor after
configuring the serial port of the computer. The serial communication may be set as
follow: 8 data bits; 1 stop bit; no parity bit; no flow control; 9600 baud rate. We
write the program in the high levels language. The function is to send a character
from the PC to the microprocessor. The microprocessor will in turn reply, with a
character. The character is to be input from the PC to the microprocessor through
the hyper terminal. It will receive and return a 'character+1'. Example, you send an
'A' to the microprocessor, the hyper terminal will receive a 'B'. If you send a '1',
you receive a '2'.

```
----------------------------------------------------
```
P4.17 PC Communication With MicroP
```
----------------------------------------------------
```

```
#include "startup.bas"
        byte  recu_pc
        initrs232in()
        Do                     'loop here
             If recu_pc<>0 Then 'character from PC
             Putcharrs232(recu_pc+1)    'send back to PC from MicroP
             Recu_pc=0
             End If
        Loop
Function putcharrs232(x)    'send a character back to PC
        Do
        loop until SCSR.7=1
        SCDR=x
        End Function
Function initrs232in()         'rs232 initialisation
        ASM  ldx  #rs232in 'activate interrupt rs232
        ASM  stx  $00FE
        End Function
'Character send to microprocessor
        ASM rs232in: staa recu_pc
        ASM rti
```

Chapter 5
Electronics Control

5.1 Servo Motor Control

Laboratory experiment for the servo motor is set up to demonstrate and study the characteristics of the implemented PID controller system. The motor's controller output is affected by the backslash. We eliminate the backslash by using the backslashing gear for the dc motor. The gearbox of the motor can be of any selection ratio for use. The joystick is used to jack in the input voltage reference signal to the motor plant. The selectable switches are to select the relevant P, PI, PD and PID controllers for analyses. We connect the controller output to the motor. The BNC connector for the servo motor locates at the right-hand side of the controller box. We can change each of the proportional, integrator and differentiator parameters by the parameters' values selection switches. The meters gives the readout of the analog voltage readings. The actual readout connects to the oscilloscope through another BNC connector located at the bottom right of the control box. From here, we study the transient and steady state characteristics of the motor plant. Instead of manual input from the joystick, we can adjust the black knob to input the ranges of frequencies to the servo motor. A square wave signal oscillator is developed to eliminate the input signal required from the function generator. Hence, we save space from the occupied function generator (Figs. 5.1 and 5.2).

© Springer Science+Business Media Singapore 2016
T.S. Ng, *Real Time Control Engineering*,
Studies in Systems, Decision and Control 65,
DOI 10.1007/978-981-10-1509-0_5

Fig. 5.1 Servomotor
controller box

Fig. 5.2 Error detecting
circuit

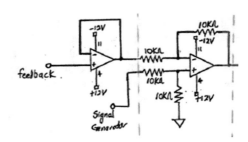

   As shown in the error detecting circuit, the voltage follower serves the purpose
of stabilizing the feedback signal from the motor plant. Following the differentiator
circuit to produce the error signal between the voltage signal feedback and the
desired input voltage signal from the square wave generator. This error signal then
feeds back into the PID controller circuit. We observe a gain of 1 in the differentiator circuit. Figure 5.3 illustrates the full control circuit for the motor plant.

5.2   Square Wave Generator

A square wave oscillator provides the input of the controller circuit with a desired
frequency of oscillation to study the motor. The circuit according to Fig. 5.4 is the
design.

Oscillatory Circuit

Desired criterior 0.2 Hz, 1 V output:
Using 12 V d.c. power supply
Choose C = 3.3 μF; R = 0.68 MΩ; R1 = R2 = 1 MΩ; R3 = 1 kΩ; R4 = 28 kΩ

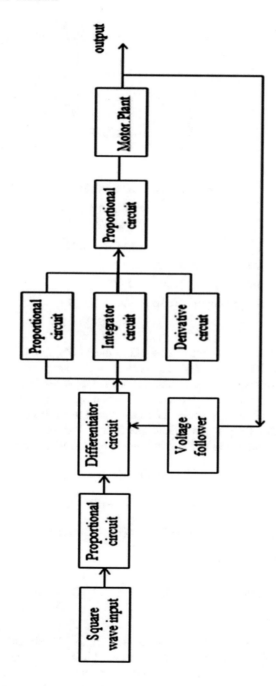

Fig. 5.3 Servomotor control block diagram

Fig. 5.4 Square wave
oscillator circuit

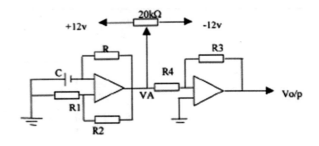

$$\text{Formula: } T = 2RC\,\text{In}[(2R1+R2)/R2]$$
$$= 2RC\,\text{In}\,3$$
$$= 4.95\ s$$
$$\text{Freq} = 1/T = 0.2\ \text{Hz}$$
$$\text{Measured value: VA} = 28\ \text{V}$$
$$\text{Amplifier Ratio : Ratio} = R3/R4$$
$$\text{Voltage Output : Vo/p} = \text{VA}.(-R3/R4)$$
$$= 28(1/28) = 1\ \text{V}$$

(5.1)

The VA value is adjustable to 12.64 kΩ to the +12 V side of the potentiometer to give a 28 V peak to peak signal. The peak to peak voltage output is equivalent to 1 V signal. Alternatively, we can generate square wave by using 555 timers. The 555 timer generates square wave oscillations when operating in astable mode. With the values of R1, R2 and C given, the frequency will range from a few Hz to several hundreds of kHz. To provide a low frequency, you can replace the 0.01 μF capacitor C with a higher value. The formula to calculate the frequency is given by:

$$T_L = 0.693 * R2 * C \qquad\qquad (5.2)$$

$$T_H = 0.693 * (R1 + R2) * C \qquad\qquad (5.3)$$

$$1/f = 0.693 * C * (R1 + 2 * R2) \qquad\qquad (5.4)$$

The duty cycle is given by:

$$\% \text{ duty cycle} = 100 * (R1 + R2)/(R1 + 2 * R2)$$
$$= 50\%$$

(5.5)

For ensuring an approximately 50 % duty ratio, R1 should be small when compared to R2. But R1 should be no lesser than 1 kΩ. A good choice would be R1 in kilo-ohms and R2 in mega ohms. You can then select C to tune the range of desired frequencies. With values of R1 = 100 kΩ; R2 = 2 MΩ; C = 1 μF; RL = 1 kΩ. for Eq. (5.4), the frequency of 0.35 Hz can be found. However,

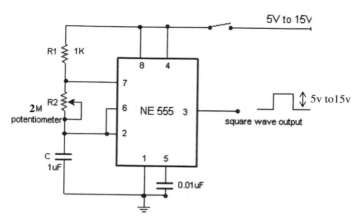

Fig. 5.5 Timer 555 square wave generator

by changing the resistor R1 to 1 kΩ, it produces a more precise square wave of 50 % duty cycle. Thus, R1 is the compromising factor leading to the 50 % duty cycle, square wave generator. We can connect a load resistor RL across pin 8 and pin 3 of the timer as a pull-up resistor. By replacing the adjustable R2 resistor with a diode (IN914) connected in parallel with a fixed R2 (anode of diode to pin7, cathode to pin6 of timer), we can have a similar effect as a timer oscillator circuit. The fixed R2 resistor (parallel with the diode), can be of equal value as R1. The output is equal to the supply voltage (Fig. 5.5).

5.3   PID Controller

The proportional-integral-derivative controller is used widely in the industries. The direct and easy implementation has made it popular in the modern control system. Assuming simple amplifiers connected in RC forms as shown in Fig. 5.6. The final gain for the PID circuit is derived as follow, where we set Kp gain to 1.

$$
\begin{aligned}
G &= kp(1/Tis + Tds + Kp)\\
&= Ki/s + KDs + K; \quad \text{where} \quad Ki = kp/Ti, \quad KD = kp * Td
\end{aligned}
\tag{5.6}
$$

Figure 5.7 shows the full implementation of the analogue PID controller. The settings are as follow: CDi = 0.1 μF; CIf = 0.22 μF; C4 = 10 nF

R1 = 1.5 kΩ;   R2 = 15 kΩ;   R3 = 4.7 kΩ;   Rkf = 10 kΩ;   Rki = 10 kΩ;

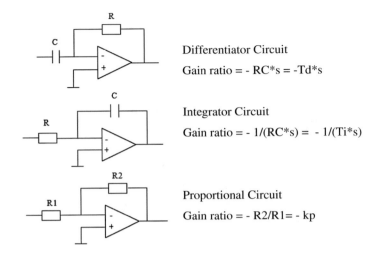

Fig. 5.6 Differentiator, integrator, proportional circuits

We define integrator, differentiator and summer corner frequencies as:

fi1 = 0.2 Hz;   fi2 = 4 Hz;   fd1 = 90 Hz;   fd2 = 800 Hz;   fs = 1 kHz

Resistor calculated for the summer corner frequency is:

$$R4 = 1/(2\pi * C4 * fs) = 15.92 \text{ k}\Omega \quad \text{(choose } R4 = 16 \text{ k}\Omega)$$

Proportional d.c gain is, $K = Kp * kp = (Rkf * R4)/(Rki * R3) = 3.404$
Suitable resistor for integrator corner frequencies are:

$$RIf = 1/(2\pi * CIf * fi1) = 3.617 \text{ M}\Omega \quad \text{(choose } RIf = 3.6 \text{ M}\Omega)$$
$$RIi = 1/(2\pi * CIf * fi2) = 0.1809 \text{ M}\Omega \quad \text{(choose } RIi = 180 \text{ k}\Omega)$$

Proportional d.c gain is, $Ki = (RIf * R4)/(RIi * R2) = 21.33$ Suitable resistor for differentiator corner frequencies are:

$$RDf = 1/(2\pi * CDi * fd1); \quad RDf = 17.68 \text{ k}\Omega \quad \text{(choose } RDf = 18 \text{ k}\Omega)$$
$$RDi = 1/(2\pi * CDi * fd2); \quad RDi = 1.989 \text{ k}\Omega \quad \text{(choose } RDi = 2 \text{ k}\Omega)$$

Proportional d.c gain is, $KD = (RDf * R4)/(RDi * R1) = 96$

$$S(s) = R4/C4 = R4/(R4.C4.s + 1)$$

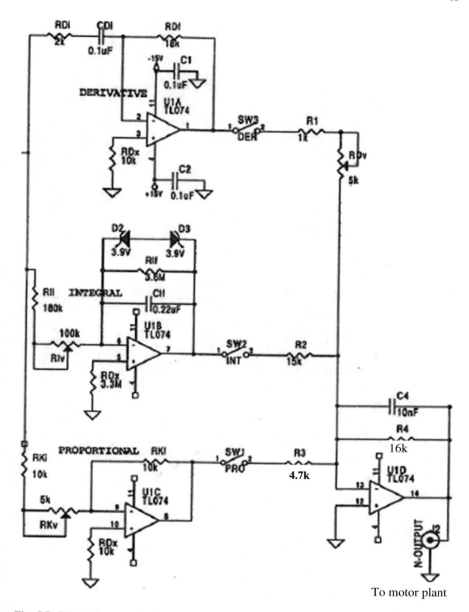

Fig. 5.7 PID analogue controller

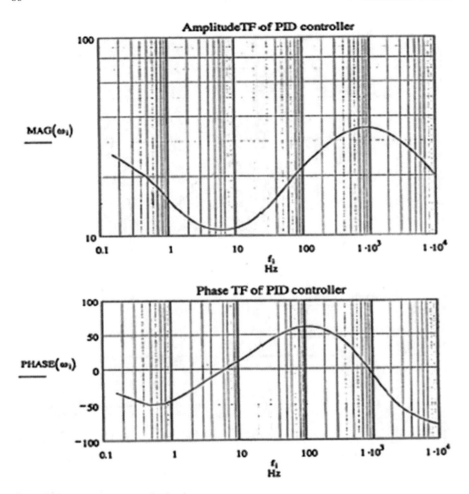

Fig. 5.8 Frequency response bode plot

$$G(s) := \left[\frac{R_{Kf}}{R_{Ki}} \cdot \frac{R_4 \cdot S(s)}{R_3} + \frac{R_{If}}{R_{Ii} \cdot (1 + s \cdot R_{If} \cdot C_{If})} \cdot \frac{R_4 S(s)}{R_2} + \frac{s \cdot C_{Di} \cdot R_{Df}}{1 + s \cdot C_{Di} \cdot R_{Di}} \cdot \frac{R_4 \cdot S(s)}{R_1} \right]$$

$$(5.7)$$

The response Mag db(w) = 20 log (|G(jw)|); N = 60,000, i = 1 to N Phase angle (w) = arg(G(jw))180/π; wi = i; fi = wi/2π (Fig. 5.8).

5.4   Control of an Electro-pneumatic Mechanism

The analogue voltage control circuit measures the differences between the desired set voltage and the feedback voltage to operate the solenoid valve of a system. The electronics circuit diagram is as shown in Fig. 5.9. The function of the first amplifier in the circuit is to calibrate the errors of the transducer signal feedback. We define this as the zeroing of the voltage feedback to a reference point. Next, we can set the desired voltage at the terminal of the differential amplifier. The voltage signal will be fed into the voltage comparator together with the feedback voltage. A supply voltage of 5 V offsets the differences of the set voltage and the feedback voltage in the next amplifier circuit. Finally, the output of the analogue control circuit is fed forward to the solenoid valve of the rodless cylinder to initiate the linear motion. The electronic circuit of the analogue voltage controller consists of three types of amplifier circuits as shown above. We can ground the inverting amplifier using a 1 kΩ, R3 resistor (Fig. 5.10).

$$\text{For differential amplifier,}\quad \text{Vo} = \frac{R2}{R1}(V2 - V1) \tag{5.8}$$

$$\text{For inverting amplifier,}\quad \text{Vo} = -\frac{R2}{R1}V1 \tag{5.9}$$

$$\text{For summing amplifier,}\quad \text{Vo} = -\left(\frac{R3}{R1}V1 + \frac{R3}{R2}V2\right) \tag{5.10}$$

The analogue signal control circuit output voltages for each of the stages are derived as follow. For simplicity, we use 1 kΩ value for all the resistors. So the amplification is straight forward.

$$\text{First Stage:}\quad V1 = -\left(\frac{R3}{R1}V\text{trim} + \frac{R3}{R2}V\text{in}\right) \tag{5.11}$$

$$\text{Second Stage:}\quad V2 = -\frac{R6}{R4}V1 \tag{5.12}$$

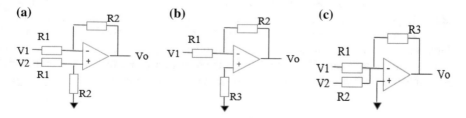

Fig. 5.9 Amplifier circuits **a** Differential. **b** Invertor. **c** Summing

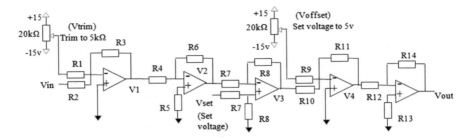

Fig. 5.10 Analogue signal control circuit

$$\text{Third Stage:} \quad V3 = \frac{R8}{R7} = (Vset - V2) \tag{5.13}$$

$$\text{Fourth Stage:} \quad V4 = -\left(\frac{R11}{R9}Voffset + \frac{R11}{R10}V3\right) \tag{5.14}$$

$$\text{Final Stage:} \quad Vout = -\left(\frac{R14}{R12}\right)V4 \tag{5.15}$$

The voltage input to the displacement output of the rodless cylinder mechanism system is as shown in the plot. The static sensitivity is calculated approximately as 5.625 V/m. There is an offset in the initial system as shown in the graph. That arises the need for the 20 kΩ potentiometer. The zeroing adjustment of the transducer potentiometer is to eliminate the error of the feedback voltage, from the displacement transducer when starting. The 20 kΩ referencing potentiometer calibrates the misalignment of the signal feedback, due to some physical, mechanical error or degradation of the components which inherits in the system. We can adjust the potentiometer or trimmer to counteract on the positive or negative transducer signal feedback. The system is trimmed to 5 kΩ as shown in the circuit. The position of the adjusted voltage of the trimmer can eliminate the error of the transducer feedback voltage (as referenced to the static sensitivity graph). For example, when the set and offset voltages are all set to zero, there is no distance travel or any displacement. So the voltage from the transducer should feedback a 0 V signal. But when referring to the static sensitivity graph, there is an offset of 1 V in the system for zeros displacement. So we trim the voltage trimmer to a negative 1 V same signal. It eliminates the transducer signal, so the final output of the analogue control circuit is zero at the beginning. Therefore, there will be no displacement for the linear system before we inject a desired set voltage (Fig. 5.11).

Also, the voltage offset is set to 5 V for the system because the rodless cylinder mechanism is referenced and positioned to this fixed distance at the beginning.

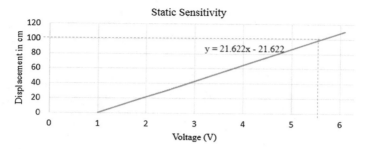

Fig. 5.11 Static sensitivity graph

As shown in the graph, the distance for the 5 V offset will be approximately 89 cm away from the starting end of the rodless linear mechanism system. Therefore, the two potentiometers serve as a signal compensation and voltage allowance for the mechanical system.

Chapter 6
Electrical System

6.1 Elevator Control

PLC control systems are invested heavily in the market to perform many different control functions. Brands like Hitachi, Omron, Allen Bradley, etcs, are popular in their usage, serving the various sectors in the industrial. Due to ease of maintenance and the flexibility in configuration, the PLC control system is used worldwide. Elevator control employs PLC.

6.2 Programmable Logic Controller

We used the Hitachi E-Series PLC for a large number of sequential control applications. The ladder logic diagram is programmed and applied in many applications. The operation of the elevator control for a four storey building is one example. The elevator or lift system consists of three PLC program structures. They are the lift up/down control structure, the lift door open/close control structure and the lift lighting control structure. Figure 6.1 is the lift system layout for the design. Table 6.1 shows the standard input and output lift functions for the operational control of the lift system. There are altogether thirteen push buttons, eight limit switches and seventeen indicating lights in the elevator system. Besides, we have four solenoid relays to control the lift up/down and door open/close positions.

6.3 Ladder Diagram Control Structures

The ladder logic control design for the lift system consists of three main parts. They are

© Springer Science+Business Media Singapore 2016
T.S. Ng, *Real Time Control Engineering*,
Studies in Systems, Decision and Control 65,
DOI 10.1007/978-981-10-1509-0_6

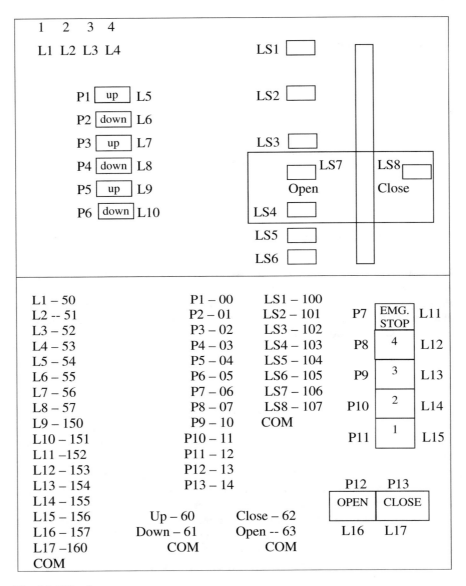

Fig. 6.1 Lift references

1. Part 1: Indicating Lights
2. Part 2: Lift door open/close
3. Part 3: Lift up/down

The structural design makes it much more easy to learn and understand. Furthermore, lift control designer can just concentrate and change the required part

I/Os		Functions
Inputs	P1 to P13	Push button switches
	LS1 to LS8	Limit switches
Outputs	L1 to L17	Indicator lights
	Up	Lift motor 'up' control
	Down	Lift motor 'down' control
	Open	Lift door 'open' control
	Close	Lift door 'close' control

Table 6.1 Elevator I/O functions

without worries or concerns of undoing the whole system. The designer can then combine the three parts as one continuous ladder control diagram for the elevator system.

6.3.1 Part 1:- Indicating Lights

As shown in Fig. 6.2 is the ladder diagram structure for the lift indicating lights. We installed these lights outside the lift for indication to passengers waiting outside the lift. We connect the internal relays 50–53 of the PLC to the indicating lights outside the elevator. The light indicators from 53 to 50 are from level 4 to level 1 respectively. LS2 activates the 4th storey limit switch, LS3 represents the 3rd storey limit switch and so on. We used only four limit switches in the design.

Alternatively, we may want to use two limit switches to signal for the lift indicator (Fig. 6.3). It depends on the design engineer whether to fix one or two

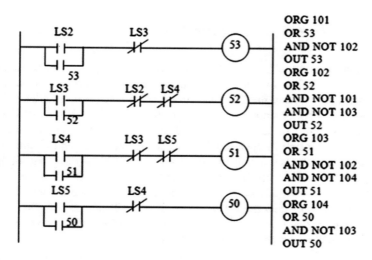

Fig. 6.2 Indicating lights

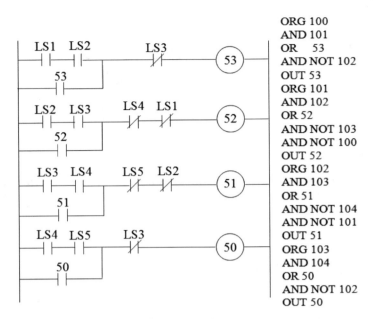

Fig. 6.3 Indicating lights alternate design

limit switches for activating the light indicators. Additional limit switches will ensure a more fully proofed system at the expenses of having more maintenance and costs. Moreover, additional hardware increases the chances of hardware failure of the system also.

6.3.2   *Part 2:- Lift Door Open/Close*

The solenoid relay 63 of Fig. 6.4 energises the lift door open motor while the coil 62 energise the lift door to close. Lamp indicators are 157 and 160. We use the on delay timer T02, to delay the lift door closing when it is open. The delay is set to 2 s. When the elevator door is opening, pressing the close button P13 will not activate the lift door to close. On the other hand, pressing the open button P12 will bypass the door closing to open the lift door. The contacts 60 and 61 are from the lift up/down motors. We install the on delay timer T04 to delay the timing for opening the lift door when the elevator reached the level or when the up/down motor has de-activated. The timer is set to hold for one second before the elevator door operates to open. So the lift up/down motor will not be activated first (6 s delay). Another scenario is to press the buttons outside the lift to be able to open the lift door. The second rung of the ladder diagram of the buttons P1 to P6 hold the lift door back from closing when the lift stations at the respective level.

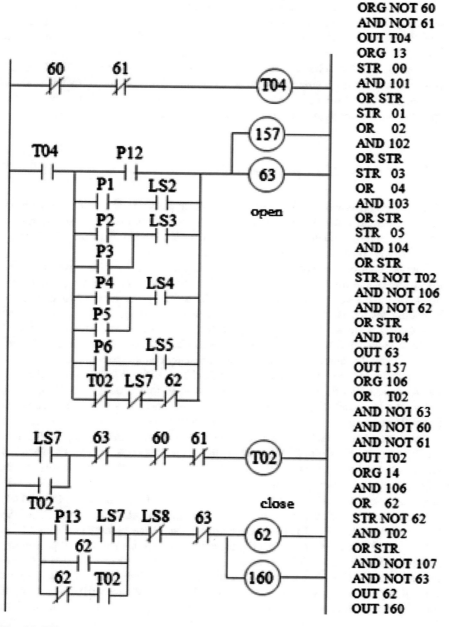

Fig. 6.4 Lift open/close

The first line in the elevator open/close ladder diagram ensures the elevator is stationary before it can open its door. While the lift is moving, the coil 60 or 61 will be activated to open its contact. Hence, logic line one of the ladder diagram is open circuit. So the lift door will not be opening. Once reached (after delay 1 s), the lift door opens until it touches the open limit switch LS7. Conversely, the lift door will close (if there is no interference) until it activates the close limit switch LS8, to de-energize the close relay motor 62. Then, the program continues to the lift up/down ladder diagram to decide about the elevator vertical movement.

6.3.3   Part 3:- Lift Up/Down

The lift up and down moving control system is shown in the ladder diagram of Fig. 6.5. Beside it are the mnemonics or logic codes for the ladder diagram program. The up/down control of the elevator is cut off by the limit switches (from LS2 to LS5) in each of the levels when activated.

The lift control is manually input by pressing any of the six buttons outside the lift or any of the four buttons inside the lift. Altogether, we have ten press button input for the elevator control system. The lift up/down movement activates when LS8 limit switch is closed. Furthermore, it depends on the location level of the elevator to decide whether to move up or down. That depends on the activation of the limit switches (LS2 to LS5) in each level. The indicating lights inside the lift, from L11 to L15 (contactor output 152 to 156) only connect to light indicators. We do not connect them to any contact point (152 to 156). Whereas, the indicating lights outside the lift, from L5 to L10 (contactor 54 to 57 and 150 to 151), all are connected to contactor points and relays. For instance, if we pressed button P9 (level 3) inside the lift, the button P2 or P3 will light up. Which button light up depends on whether the lift is above or below the level 3. Let us say, the lift is in level 2 or 1, so P3 (light L7) lights up. If the lift is at level 4, L6 will lit instead of L7 outside the lift. If we press P2 or P3 outside the lift, the P9 button inside the lift will light up also. Thus, passengers outside or inside the lift know the elevator will arrive and stop at the levels where there are lighted buttons. The relays (60 and 61) each connects to their respective motor for activating the lift up and down. The LS8 limit switch ensures the lift is closed before it can move up/down. We install on delay timer T05 to T08 to control the elevator from overshot when the passengers had pressed the lift from different levels. These four timers have a delay of 6 s each, to cater time for the lift to stop at the activated level before it responses to its next vertical (up/down) movement in the same direction. Whereas the three seconds timer T09 and the two seconds timer T11, allow enough time to open the lift door first before it responses to the next passenger level for the lift to move in the opposite direction. The delay prevents the lift from responding to move in the opposite direction when it has reached the passenger level. Timers T09 and T11 are

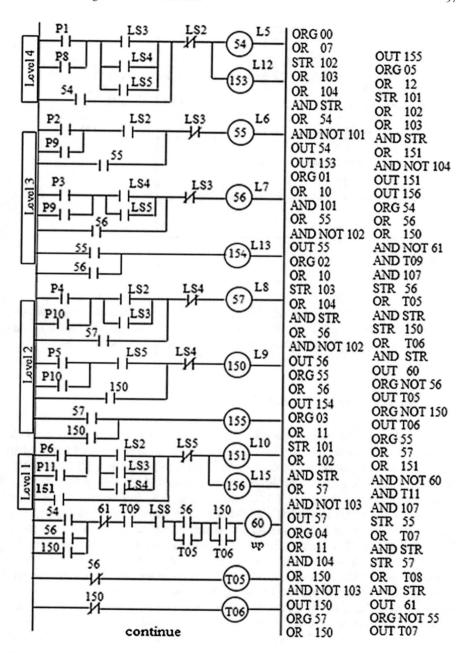

Fig. 6.5 Lift up/down

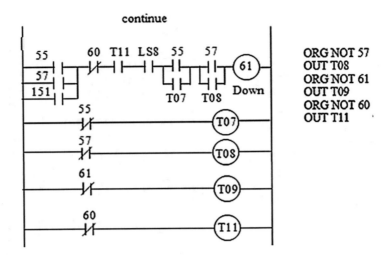

Fig. 6.5 (continued)

set differently to prevent both up/down motors of the lift from energizing simultaneously. Together with the ladder diagram controlling the lights and the lift door opening and closing, the elevator functions as a complete working system.

6.4 Safety Control Features

We can add safety features to the lift control system. For example, we can put two more limit switches to stop the lift at the highest and the lowest level. Limit switch LS1 is the safety control to halt the elevator from rising when the LS2 limit switch at level 4 fails. Similarly, LS6 limit switch is mounted to prevent the lift from descending further if limit switch LS5 fails. Secondly, we implement a timer delay or on delay timer T01 to activate the lift to rise. The purpose is to delay energizing the upward lift motor if both the up and down motors activate at the same time when the lift door (LS8 limit switch) is at closed position. That is cause by both the upper and lower level passengers energizing the lift buttons when the lift is held stationary at the middle level (between both the passengers level). So, the lift will activate the move down motor 61 first. Furthermore, we can install the emergency button inside the lift too. Button P7 is the emergency latch button to halt the elevator at the standstill during the emergency. Another implementation is to move the elevator to the level 1 or ground level when nobody is using. If any of the

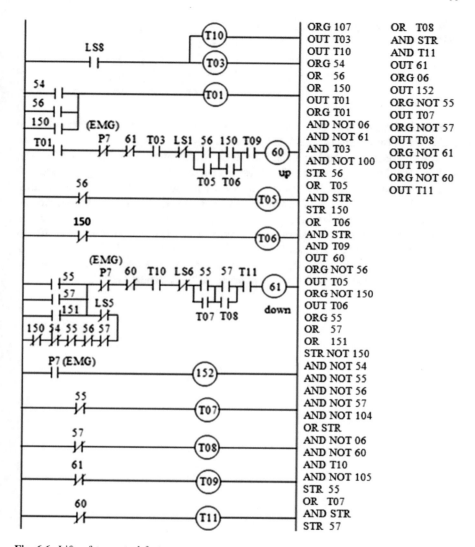

Fig. 6.6 Lift safety control features

buttons from P1 to P6 and from P8 to P11 are not press, the elevator will automatically lower itself to the ground level (level 1). That is the initial stationary position (level) of the lift to wait for passengers. Fifth, we can add on delay timer (T03) to delay the lift from moving up once the door is closed. Lastly, we add a timer delay, T10 to have different timing for energizing the lift down motor (61) once the lift is closed. The on delay timer, T03 is set to two seconds while T10 is set to 1 s delay. Thus, the timer T01 can be considered redundant for the situation. These six additional features of the lift system are as shown in Fig. 6.6.

Table 6.2 Elevator on-delay timer settings	T01	T02	T03	T04	T05 T06 T07 T08	T10	T09	T11
	1 s	2 s	2 s	1 s	6 s	1 s	3 s	2 s

We can include them to replace the last eight rungs of Fig. 6.5 to have the complete safety elevator system (Table 6.2).

The lift timing control is done by the on delay timers from T01 to T11. The settings for the timings ensure the elevator operates without fail. It guarantees the smooth transition of the lift to the next level and the efficiency of the lift control system. With the presence of the timers T10 and T03 in the system design, we can lengthen T02 to 4 s delay. The extra delay time holds the lift door longer in the open standstill position. It caters for the concern of the elderly to enter the lift. At the same time, we can shorten T05 to T08 to 4 s also. In reducing these timing delays, enable the lift passenger buttons to be more active. By doing so, we improved the efficiency and user friendliness of the elevator system.

Chapter 7
Power Flow

7.1 Power System Analysis

Numerical methods are used to solve for complex non-linear problems in power system analysis. One of the most popular techniques is the approximation of the non-linear algebraic equations to linearise equations. The Newton-Raphson method is an iterative algorithm, which only makes use of the first few terms of the Taylor series function f(x). It is used practically in load flow power network. Power efficiency is crucial for load flow transmission and distribution. Transmission lines circuit protections and overloads need to be calculated to monitor and control the network system. So network design and line regulation are optimized for better efficiency.

7.2 Newton Raphson Formulation

By the Newton-Raphson formulation, we have the first-order terms from the Taylor series expansion shown in Eq. (7.1). Where e stands for the amount of error computed. x0 is the initial guess for the unknown variable. We set f (x0 + e) to zero and solve for the error in the system (see Eq. (7.2)).

$$f(x0 + e) = f(x0) + f'(x0)e \qquad (7.1)$$

$$e0 = -f(x0)/f'(x0) \qquad (7.2)$$

$$x_{n+1} = x_n + e_n \qquad (7.3)$$

The unknown value for the system is updated as in Eq. (7.3) until the error is zero.

© Springer Science+Business Media Singapore 2016
T.S. Ng, *Real Time Control Engineering*,
Studies in Systems, Decision and Control 65,
DOI 10.1007/978-981-10-1509-0_7

We give an illustration of the N-R program written in Pascal language. It computes two simultaneous equations using Newton-Raphson formulation. The convergence is accurate. We just need to input two unknown values of the approximations as in the equations, and the program computes the result by itself (Fig. 7.1).

Program 7.1: Newton Raphson

```
Program Newton_Raphson;

VAR
X1,X2,F1,F2,X11,X22,T:REAL;
J11,J22,J12,J21,IJ11,IJ22,IJ12,IJ21,DET:REAL;
COUNT:INTEGER;
BEGIN
WRITELN;
WRITELN('Newton Raphson Technique for Simultaneous Equations');
WRITELN('     3     2     2');
WRITELN('F1= 2X1  + 3X1 X2 - X2  - 2 = 0');
WRITELN('     2     2     2');
WRITELN('F2= X1X2  + 2X1  - 3X2  + 16 = 0');
WRITELN;
WRITELN('ENTER TWO VALUE OF APPROXIMATION');
WRITELN;
READLN(X1,X2);
WRITELN('X1 = ',X1:4:5,'   X2 = ',X2:10:5);
WRITELN;
WRITELN;
WRITELN('COUNT':8,'X1':17,'X2':17);
WRITELN(COUNT:8,X1:17:5,X2:17:5);
X11:=1.0;
X22:=1.0;
```

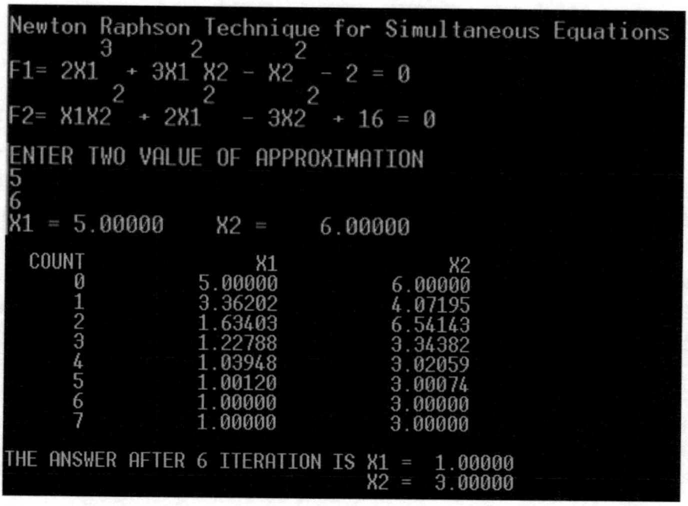

```
Newton Raphson Technique for Simultaneous Equations
F1= 2X1³ + 3X1²X2 - X2² - 2 = 0
F2= X1X2² + 2X1² - 3X2² + 16 = 0

ENTER TWO VALUE OF APPROXIMATION
5
6
X1 = 5.00000    X2 =    6.00000

    COUNT              X1                    X2
        0          5.00000               6.00000
        1          3.36202               4.07195
        2          1.63403               6.54143
        3          1.22788               3.34382
        4          1.03948               3.02059
        5          1.00120               3.00074
        6          1.00000               3.00000
        7          1.00000               3.00000

THE ANSWER AFTER 6 ITERATION IS X1 =  1.00000
                                X2 =  3.00000
```

Fig. 7.1 Result of the N-R computation

```
T:=0.001;
WHILE (ABS(X11)>T) OR (ABS(X22)>T) DO
BEGIN
  F1:=2*EXP(3*LN(X1))+3*SQR(X1)*X2-SQR(X2)-2;
  F2:=X1*SQR(X2)+2*SQR(X1)-3*SQR(X2)+16;
  J11:=6*SQR(X1)+6*X1*X2;
  J12:=3*SQR(X1)-2*X2;
  J21:=SQR(X2)+4*X1;
  J22:=2*X2*X1-6*X2;
  DET:=J11*J22-J21*J12;
  IJ11:=J22/DET;
  IJ12:=-J12/DET;
  IJ21:=-J21/DET;
  IJ22:=J11/DET;
  X11:=-(IJ11*F1+IJ12*F2);
  X22:=-(IJ21*F1+IJ22*F2);
  X1:=X1+X11;
  X2:=X2+X22;
  COUNT:=COUNT+1;
  WRITELN(COUNT:8,X1:17:5,X2:17:5);
END;
WRITELN;
WRITELN(' THE ANSWER AFTER ',COUNT-1,' ITERATION IS X1 =  ',X1:1:5);
WRITELN('                        X2 =  ',X2:1:5);
```

7.3   Load Flow Analysis Using Newton Raphson

Consider the following two-bus network system. We use the Newton-Raphson technique to investigate the reactive power compensation and solve the load flow problem. The Newton-Raphson technique is a powerful tool, which provides convergent most of the time, comparing to the Gauss-Seidel method. The power generated assumed positive, and the load is assumed negative. We can compute the voltages, angles, active and reactive power at the buses. Once the power factor and system losses are known, we can control the power system operation and enhancement for further planning of the power system network. The computation is base on the assumption of a steady-state load flow operation. Where the parameters of the A.C power line are determined.

$$\text{Pi calc} = \sum_{k=1}^{n} |V_i V_k V_{ik}| \cos(\Phi ik + \delta k - \delta i) \tag{7.4}$$

$$\text{Qi calc} = -\sum_{k=1}^{n} |V_i V_k V_{ik}| \sin(\Phi ik + \delta k - \delta i) \tag{7.5}$$

$$\Delta Pi = \text{Pi spec} - \text{Pi calc} \tag{7.6}$$

$$\Delta Qi = \text{Qi spec} - \text{Qi calc} \tag{7.7}$$

$$J = \begin{bmatrix} \dfrac{\partial P_{icalc}}{\partial \delta_i} & \dfrac{\partial P_{icalc}}{\partial |V_i|} \\ \dfrac{\partial Q_{icalc}}{\partial \delta_i} & \dfrac{\partial Q_{icalc}}{\partial |V_i|} \end{bmatrix} \tag{7.8}$$

$$\begin{bmatrix} \Delta P_i \\ \Delta Q_i \end{bmatrix} = \begin{bmatrix} \dfrac{\partial P_{icalc}}{\partial \delta_i} & \dfrac{\partial P_{icalc}}{\partial |V_i|} \\ \dfrac{\partial Q_{icalc}}{\partial \delta_i} & \dfrac{\partial Q_{icalc}}{\partial |V_i|} \end{bmatrix} \cdot \begin{bmatrix} \Delta \delta_i \\ \Delta V_i \end{bmatrix} \tag{7.9}$$

$$\begin{bmatrix} \Delta \delta_i \\ \Delta V_i \end{bmatrix} = [J]^{-1} \begin{bmatrix} \Delta P_i \\ \Delta Q_i \end{bmatrix} \tag{7.10}$$

$$\begin{bmatrix} \delta_i^{(r+1)} \\ V_i^{(r+1)} \end{bmatrix} = \begin{bmatrix} \delta_i^r \\ V_i^r \end{bmatrix} + \begin{bmatrix} \Delta \delta_i^r \\ \Delta V_i^r \end{bmatrix} \tag{7.11}$$

Expression (7.4) and (7.5) shows the calculated active and reactive power at bus i. The differences between the specifications and the calculated power in Eqs. (7.6) and (7.7) are the changes in the active and reactive power. The Jacobian matrix of Eq. (7.8) is the partial derivative of Pi and Qi. Equation (7.9) is the system equation. By substituting the Jacobian matrix into the system, we derive Eq. (7.9).

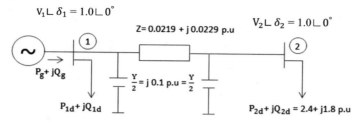

$V_1 L \delta_1 = 1.0 L 0°$

$Z = 0.0219 + j0.0229$ p.u

$V_2 L \delta_2 = 1.0 L 0°$

$P_g + jQ_g$

$\frac{Y}{2} = j0.1$ p.u $= \frac{Y}{2}$

$P_{1d} + jQ_{1d}$

$P_{2d} + jQ_{2d} = 2.4 + j1.8$ p.u

Fig. 7.2 Two-bus power network circuit

By rearrangement, we arrive at Eq. (7.10). Equation (7.11) is the updated value of the voltages and the angles. These values are then substituted back into the Pi calc and the Qi calc for another round of iteration. We matched the mismatch equations ΔPi and ΔQi against a tolerance value. The computation ended when it satisfies the tolerances specified.

We specify the voltage and angle of a slack bus to be the referencing source. It makes up for the losses between the generated power and the load power. If we know the power and the voltages unknown, it is known as the load bus. Another name for it is the PQ bus. Generator bus is where we already know the real power and the magnitude of the voltage. We need to find the reactive power and the voltage angle of the PV bus (Fig. 7.2).

$$Y11 = Y22 \tag{7.12}$$

$$Y12 = Y21 \tag{7.13}$$

$$Z12 = 0.0219 + j0.0229$$
$$= 0.031686 L 46.28 \tag{7.14}$$

$$Y12 = 21.8 - j22.8$$
$$= 31.35 L - 46.28 \tag{7.15}$$
$$= 31.55 L 133.72$$

$$Y/2 = j0.1$$
$$= 0.1 L 90 \tag{7.16}$$

$$Y11 = Yt \tag{7.17}$$

$$Yt = Y/2 + Y12$$
$$= j0.1 + 21.8 - j22.8$$
$$= 21.8 - j22.7 \tag{7.18}$$
$$= 31.47 L - 46.15$$

(Fig. 7.3).

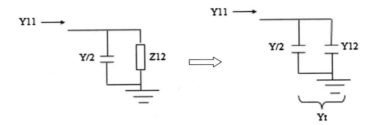

Fig. 7.3 Transmission line admittance circuit

```
Microsoft Windows [Version 6.1.7601]
Copyright (c) 2009 Microsoft Corporation.  All rights reserved.

C:\Users\admin>copy con data1.txt
31.47 -46.15
31.55 133.72
31.55 133.72
31.47 -46.15
^Z
        1 file(s) copied.

C:\Users\admin>_
```

Fig. 7.4 DOS environment (Y-parameters input)

Go to the DOS environment to key in the Y-parameters' data for calculations. Hit Ctrl-Z (return) to save the file (Fig. 7.4).

Let us write a program using Pascal to determine the load flow problem for the following 3 cases.

Case 1: Determine the voltage and its phase angle at bus 2 (V_2 $∠_2$).
Case 2: Determine the rating of the capacitor (Q_c) that if connected at bus 2, it would make $V_1 = V_2$ provided that the generator limits are not being exceeded; Pg = 3.0 PU; 2.5 ≥ Qg ≥ −1.5 PU
          Hint: If you exceed the limits, you can use an inductor (Q_L) at bus 1.
Case 3: Determine the voltage at bus 2 when the load is suddenly switched off with the capacitor at bus 2 and the inductor at bus 1, remain connected.

For all cases, calculate the active and reactive power generated at bus 1 (P_g and Q_g). Determine the system losses and power factor. Use a tolerance of 0.0001. Output the results showing all the power and voltages of both buses.

The algorithm for the program written is as shown at the end of the chapter. The input guessing values for convergence determines the number of iterations with a given tolerance. The convergence is best for the approximated X2 (radian) value of zero. For case 1 the limited X2 (radian) value is not exceeding 0.81. For case 3, the maximum for X2 (radian) is not exceeding 0.71. The V2 magnitude approximation can be of any positive value not less than 1. For case 2, the convergent for the NR algorithm is between 14.33 and −5.7 for Q2 demand value input estimation. The X2 (radian) angular estimation is best at zero.

Figures 7.5, 7.6 and 7.7 proved the results for the computation of the written program for the 3 cases.

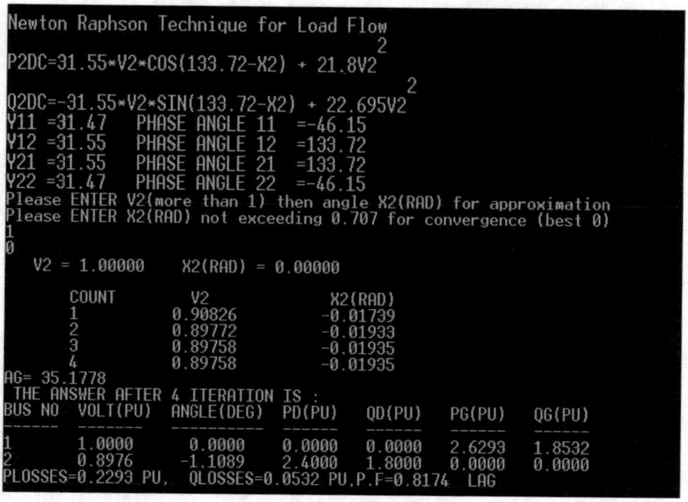

```
Newton Raphson Technique for Load Flow
                               2
P2DC=31.55*V2*COS(133.72-X2) + 21.8V2
                                    2
Q2DC=-31.55*V2*SIN(133.72-X2) + 22.695V2
Y11 =31.47    PHASE ANGLE 11  =-46.15
Y12 =31.55    PHASE ANGLE 12  =133.72
Y21 =31.55    PHASE ANGLE 21  =133.72
Y22 =31.47    PHASE ANGLE 22  =-46.15
Please ENTER V2(more than 1) then angle X2(RAD) for approximation
Please ENTER X2(RAD) not exceeding 0.707 for convergence (best 0)
1
0
   V2 = 1.00000    X2(RAD) = 0.00000

      COUNT          V2              X2(RAD)
        1          0.90826         -0.01739
        2          0.89772         -0.01933
        3          0.89758         -0.01935
        4          0.89758         -0.01935
AG= 35.1778
 THE ANSWER AFTER 4 ITERATION IS :
BUS NO  VOLT(PU)  ANGLE(DEG)  PD(PU)   QD(PU)   PG(PU)   QG(PU)
------  --------  ----------  ------   ------   ------   ------
1       1.0000     0.0000     0.0000   0.0000   2.6293   1.8532
2       0.8976    -1.1089     2.4000   1.8000   0.0000   0.0000
PLOSSES=0.2293 PU,  QLOSSES=0.0532 PU,P.F=0.8174  LAG
```

Fig. 7.5 Case 1 output

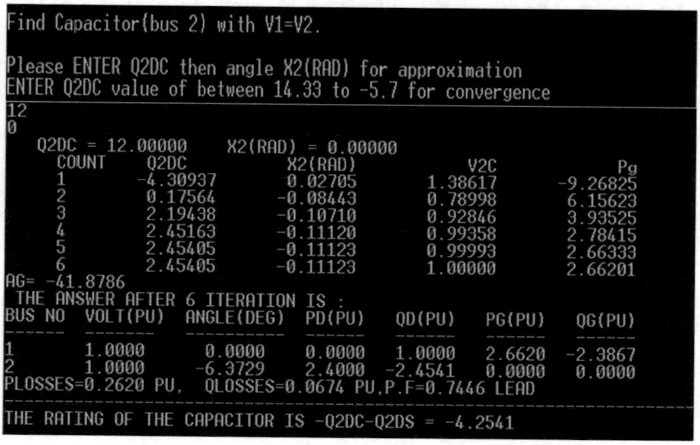

```
Find Capacitor(bus 2) with V1=V2.

Please ENTER Q2DC then angle X2(RAD) for approximation
ENTER Q2DC value of between 14.33 to -5.7 for convergence
12
0
   Q2DC = 12.00000    X2(RAD) = 0.00000
     COUNT      Q2DC          X2(RAD)         V2C            Pg
       1       -4.30937       0.02705      1.38617       -9.26825
       2        0.17564      -0.08443      0.78998        6.15623
       3        2.19438      -0.10710      0.92846        3.93525
       4        2.45163      -0.11120      0.99358        2.78415
       5        2.45405      -0.11123      0.99993        2.66333
       6        2.45405      -0.11123      1.00000        2.66201
AG= -41.8786
 THE ANSWER AFTER 6 ITERATION IS :
BUS NO  VOLT(PU)  ANGLE(DEG)  PD(PU)   QD(PU)   PG(PU)   QG(PU)
------  --------  ----------  ------   ------   ------   ------
1       1.0000     0.0000     0.0000   1.0000   2.6620  -2.3867
2       1.0000    -6.3729     2.4000  -2.4541   0.0000   0.0000
PLOSSES=0.2620 PU,   QLOSSES=0.0674 PU,P.F=0.7446 LEAD
------------------------------------------------------
THE RATING OF THE CAPACITOR IS -Q2DC-Q2DS = -4.2541
```

Fig. 7.6 Case 2 output (without inductor)

(a)

```
Find Capacitor(bus 2) with V1=V2.

Please ENTER Q2DC then angle X2(RAD) for approximation
ENTER Q2DC value of between 14.33 to -5.7 for convergence
1
0
   Q2DC = 1.00000    X2(RAD) = 0.00000
      COUNT     Q2DC          X2(RAD)              V2C            Pg
        1      2.08885       -0.09839           1.04682        1.43048
        2      2.44981       -0.11116           0.99774        2.70351
AG= -26.1697
 THE ANSWER AFTER 2 ITERATION IS :
BUS NO  VOLT(PU)  ANGLE(DEG)  PD(PU)   QD(PU)   PG(PU)   QG(PU)
------  --------  ----------  ------   ------   ------   ------
1        1.0000     0.0000    0.0000   1.0000   2.7035  -1.3285
2        1.0000    -6.3687    2.1825  -2.4498   0.0000   0.0000
PLOSSES=0.5210 PU.  QLOSSES=0.1213 PU.P.F=0.8975 LEAD

THE RATING OF THE CAPACITOR IS -Q2DC-Q2DS = -4.2498
```

(b)

```
Find Capacitor(bus 2) with V1=V2.

Please ENTER Q2DC then angle X2(RAD) for approximation
ENTER Q2DC value of between 14.33 to -5.7 for convergence
1
0
   Q2DC = 1.00000    X2(RAD) = 0.00000
      COUNT     Q2DC          X2(RAD)              V2C            Pg
        1      2.08885       -0.09839           1.04682        1.43048
        2      2.44981       -0.11116           0.99774        2.70351
        3      2.45405       -0.11123           0.99990        2.66389
AG= -27.4575
 THE ANSWER AFTER 3 ITERATION IS :
BUS NO  VOLT(PU)  ANGLE(DEG)  PD(PU)   QD(PU)   PG(PU)   QG(PU)
------  --------  ----------  ------   ------   ------   ------
1        1.0000     0.0000    0.0000   1.0000   2.6639  -1.3842
2        1.0000    -6.3729    2.4004  -2.4540   0.0000   0.0000
PLOSSES=0.2635 PU.  QLOSSES=0.0698 PU.P.F=0.8874 LEAD

THE RATING OF THE CAPACITOR IS -Q2DC-Q2DS = -4.2540
```

Fig. 7.7 **a** Case 2 output, **b** case 2 output, **c** case 2 output, **d** case 2 output

For case 2, we need to derive the subject of interest which is the ΔQ and $\Delta \delta$. The below equations' formulations result in the matrix equation of interest (Eq. 7.26) required for the computation. The first result is as shown in Fig. 7.6 when the inductor is not present to ensure we do not exceed the generator (Qg) limitation requirements.

$$\begin{bmatrix} \Delta P \\ \Delta Q \end{bmatrix} = \begin{bmatrix} J11 & J12 \\ J21 & J22 \end{bmatrix} \begin{bmatrix} \Delta \delta \\ \Delta V \end{bmatrix} \tag{7.19}$$

$$\Delta P = J11\,\Delta \delta + J12\,\Delta V \tag{7.20}$$

$$\Delta Q = J21\,\Delta \delta + J22\,\Delta V \tag{7.21}$$

$$J11\,\Delta \delta = \Delta P - J12\,\Delta V \tag{7.22}$$

(c)

```
Find Capacitor(bus 2) with V1=V2.

Please ENTER Q2DC then angle X2(RAD) for approximation
ENTER Q2DC value of between 14.33 to -5.7 for convergence
1
0
   Q2DC = 1.00000     X2(RAD) = 0.00000
      COUNT    Q2DC           X2(RAD)            V2C              Pg
        1      2.08885       -0.09839          1.04682          1.43048
        2      2.44981       -0.11116          0.99774          2.70351
        3      2.45405       -0.11123          0.99990          2.66389
        4      2.45405       -0.11123          1.00000          2.66201
AG= -27.5159
 THE ANSWER AFTER 4 ITERATION IS :
BUS NO  VOLT(PU)   ANGLE(DEG)  PD(PU)    QD(PU)    PG(PU)    QG(PU)
------  --------   ----------  ------    ------    ------    ------
1       1.0000      0.0000     0.0000    1.0000    2.6620   -1.3867
2       1.0000     -6.3729     2.4000   -2.4541    0.0000    0.0000
PLOSSES=0.2620 PU.   QLOSSES=0.0674 PU.P.F=0.8869 LEAD
----------------------------------------------------------------
THE RATING OF THE CAPACITOR IS -Q2DC-Q2DS = -4.2541
```

(d)

```
Find Capacitor(bus 2) with V1=V2.

Please ENTER Q2DC then angle X2(RAD) for approximation
ENTER Q2DC value of between 14.33 to -5.7 for convergence
14
0
   Q2DC = 14.00000     X2(RAD) = 0.00000
      COUNT    Q2DC           X2(RAD)            V2C              Pg
        1     -6.04377        0.05268          1.43470        -11.16168
        2     -3.66660        0.00642          0.63365          7.89218
        3      0.62248       -0.08869          0.81187          5.80772
        4      2.29635       -0.10884          0.94471          3.66384
        5      2.45313       -0.11122          0.99600          2.73831
        6      2.45405       -0.11123          0.99997          2.66250
        7      2.45405       -0.11123          1.00000          2.66201
AG= -27.5159
 THE ANSWER AFTER 7 ITERATION IS :
BUS NO  VOLT(PU)   ANGLE(DEG)  PD(PU)    QD(PU)    PG(PU)    QG(PU)
------  --------   ----------  ------    ------    ------    ------
1       1.0000      0.0000     0.0000    1.0000    2.6620   -1.3867
2       1.0000     -6.3729     2.4000   -2.4541    0.0000    0.0000
PLOSSES=0.2620 PU.   QLOSSES=0.0674 PU.P.F=0.8869 LEAD
----------------------------------------------------------------
THE RATING OF THE CAPACITOR IS -Q2DC-Q2DS = -4.2541
```

Fig. 7.7 (continued)

$$\Delta\delta = \frac{1}{J11}\Delta P - \frac{J12}{J11}\Delta V \tag{7.23}$$

$$\Delta Q = J21\left(\frac{\Delta P}{J11} - \frac{J12\Delta V}{J11}\right) + J22\Delta V \tag{7.24}$$

$$\Delta Q = \frac{J21}{J11}\Delta P + \left(J22 - \frac{J21J12}{J11}\right)\Delta V \tag{7.25}$$

```
Load Switched off with Capacitor(bus 2) and Inductor(bus 1) Connected.

Please ENTER V2(more than 1) then angle X2(RAD) for approximation
Please ENTER X2(RAD) not exceeding 0.707 for convergence (best 0)
44
0
   V2 = 44.00000     X2(RAD)  = 0.00000

       COUNT          V2              X2(RAD)
         1          22.25465        -0.00327
         2          11.38597        -0.00696
         3           5.95963        -0.01271
         4           3.26232        -0.02267
         5           1.94468        -0.03915
         6           1.34291        -0.06173
         7           1.12716        -0.08145
         8           1.08963        -0.08788
         9           1.08841        -0.08820
        10           1.08841        -0.08820
AG= -83.6525
 THE ANSWER AFTER 10 ITERATION IS :
BUS NO   VOLT(PU)   ANGLE(DEG)  PD(PU)    QD(PU)    PG(PU)    QG(PU)
------   --------   ---------   ------    ------    ------    ------
1        1.0000     0.0000      0.0000    1.0000    0.3468   -3.1171
2        1.0884    -5.0533     -0.0000   -4.2541    0.0000    0.0000
PLOSSES=0.3468 PU,  QLOSSES=0.1369 PU,P.F=0.1106 LEAD
```

Fig. 7.8 Case 3 output

$$\begin{bmatrix} \Delta\delta \\ \Delta Q \end{bmatrix} = \begin{bmatrix} \frac{1}{J11} & \frac{J12}{J11} \\ \frac{J21}{J11} & J22 - \frac{J21J12}{J11} \end{bmatrix} \begin{bmatrix} \Delta P \\ \Delta V \end{bmatrix} \qquad (7.26)$$

We insert an inductor (Q_L) at bus 1 because from Fig. 7.6 we had exceeded the limitations for the QG(PU) value. Figure 7.7 shows the result is within the limits after the insertion of the inductor at bus 1. When using the generator limits Qg as the conversion requirement, the active power at bus 2 is out of the range (see Fig. 7.7a). When using termination tolerances of $\Delta V22$ and $\Delta X22$, the convergence is slightly out of the range for load active power load (P2D) at bus bar 2 (see Fig. 7.7b). From the findings, the convergence requirement is the best when using changes in Pi and Qi as the terminating tolerances. (The active power load on bus 2 is exactly the same given value when the iteration stops.) (see Fig. 7.7c, d) Figure 7.8 shows the result of the voltage at bus 2 when the load is suddenly switched off with the capacitor at bus 2 and the inductor at bus 1, remain connected.

```
-----------------------------------------------------
```
Program 7.2:Newton Raphson Load Flow
```
-----------------------------------------------------
```

```
Program Newton_Raphson_Load_Flow;
VAR
Y,PHI:ARRAY[1..2,1..2] OF REAL;
READFILE : TEXT;
I,K,COUNT,NUMBER :INTEGER; STRIN:STRING;
CP2,CQ2,P2DC,Q2DC,PL,QL,X22,V22,a,b,c:REAL;
PF,CP1,CQ1,P1DC,Q1DC,P2G,Q2G,PGS,QGS:REAL;
P1DS,Q1DS,P2DS,Q2DS,PGC,QGC,AG,Q2C,Q1I,QGT:REAL;
J11,J22,J12,J21,IJ11,IJ22,IJ12,IJ21:REAL;
J1,J2,J3,J4,T,DET,V1,X1,V2,X2,V2C:REAL;
BEGIN
WRITELN('Newton Raphson Technique for Load Flow');
WRITELN('                        2');
WRITELN('P2DC=31.55*V2*COS(133.72-X2) + 21.8V2 ');
WRITELN('                        2');
WRITELN('Q2DC=-31.55*V2*SIN(133.72-X2) + 22.695V2 ');
ASSIGN(READFILE,'C:DATA1.TXT');
RESET(READFILE);
FOR I:=1 TO 2 DO
  FOR K:=1 TO 2 DO
    READLN(READFILE,Y[I][K],PHI[I][K]);
CLOSE(READFILE);
FOR I:=1 TO 2 DO
  FOR K:=1 TO 2 DO
      BEGIN
WRITELN('Y',I,K,' =',Y[I,K]:4:2,'  PHASE ANGLE ',I,K,' =',PHI[I,K]:5:2);
      PHI[I,K]:=PHI[I,K]*PI/180;
      END;
NUMBER:=0;
WHILE NUMBER<3 DO
BEGIN
CP2:=1.0;
CQ2:=1.0;
T:=0.0001;
COUNT:=0;
CP1:=0;
CQ1:=0;P1DS:=0;Q1DS:=0;
PGS:=0;P2G:=0;P2DS:=2.4;P1DC:=0;
QGS:=0;Q2G:=0;Q2DS:=1.8;Q1DC:=0;
V1:=1; X1:=0;
IF NUMBER=2 THEN
BEGIN
```

```
        Q1DS:=Q1I;
        P2DS:=P2DS-P2DS;
        Q2DS:=Q2DS-Q2DS+Q2C;
WRITELN('Load Swtched off with Capacitor(bus 2) and Inductor(bus 1) Connected.');
        WRITELN;
        END;
WRITELN('Please ENTER V2(more than 1) then angle X2(RAD) for approximation');
WRITELN('Please ENTER X2(RAD) not exceeding 0.707 for convergence (best 0)');
        READLN(V2,X2); {*not >0.81 for question 1*}
        WRITELN('   V2 = ',V2:4:5,'   X2(RAD) = ',X2:4:5);
        WRITELN;        {*not >0.74 for question 3*}
        WRITELN('COUNT':12,'V2':10,'X2(RAD)':20);
        WHILE (ABS(CP2)>T) OR (ABS(CQ2)>T) DO
        BEGIN
P2DC:=Y[2,2]*SQR(V2)*COS(PHI[2,2])+(Y[2,1]*COS(PHI[2,1]+X1-X2)*V2*V1);
Q2DC:=-Y[2,2]*SQR(V2)*SIN(PHI[2,2])-(Y[2,1]*SIN(PHI[2,1]+X1-X2)*V2*V1);
        CP2:=-P2DS-P2DC;
        CQ2:=-Q2DS-Q2DC;
        J11:=Y[2,1]*V1*V2*SIN(PHI[2,1]+X1-X2);
        J12:=Y[2,1]*COS(PHI[2,1]+X1-X2)*V1+Y[2,2]*2*V2*COS(PHI[2,2]);
        J21:=Y[2,1]*V1*V2*COS(PHI[2,1]+X1-X2);
        J22:=-Y[2,1]*V1*SIN(PHI[2,1]+X1-X2)-Y[2,2]*2*V2*SIN(PHI[2,2]);
        DET:=J11*J22-J21*J12;
        IJ11:=J22/DET;
        IJ12:=-J12/DET;
        IJ21:=-J21/DET;
        IJ22:=J11/DET;
        X22:=(IJ11*CP2+IJ12*CQ2);
        V22:=(IJ21*CP2+IJ22*CQ2);
        X2:=X2+X22;
        V2:=V2+V22;
        COUNT:=COUNT+1;
        WRITELN(COUNT:8,V2:17:5,X2:17:5);
        END;
PGC:=SQR(V1)*Y[1,1]*COS(PHI[1,1])+V1*V2*Y[1,2]*COS(PHI[1,2]+X2-X1);
QGC:=-SQR(V1)*Y[1,1]*SIN(PHI[1,1])-V1*V2*Y[1,2]*SIN(PHI[1,2]+X2-X1);
        IF NUMBER=2 THEN
        QGC:=QGC-(-Q1I); {Pt=Pg-Pload}
        PGS:=PGC+CP1;
        QGS:=QGC+CQ1;
        X2:=X2*180/PI;
        PL:=PGC-P2DS-P1DS;
        QL:=QGC-Q2DS-Q1DS;
        PF:=PGC/SQRT(SQR(PGC)+SQR(QGC));
        AG:=arctan(QGC/PGC)*180/PI;
```

```
                WRITELN('AG= ',AG:2:4);
                IF AG>0.00 THEN
                STRIN:='LAG'
                ELSE
                STRIN:='LEAD';
                WRITELN(' THE ANSWER AFTER ',COUNT,' ITERATION IS : ');
                WRITELN('BUS NO    VOLT(PU)    ANGLE(DEG)    PD(PU)      QD(PU)
PG(PU)  QG(PU)');
                WRITELN('------ ------- ---------- ------ ------ ------ ------');
                WRITELN('1',V1:13:4,X1:12:4,P1DS:10:4,Q1DS:9:4,PGC:9:4,QGC:9:4);
                WRITELN('2',V2:13:4,X2:12:4,-P2DC:10:4,-Q2DC:9:4,'0.0000':9,'0.0000':9);
                WRITELN('PLOSSES=',PL:2:4,'      PU,              ','QLOSSES=',QL:4:4,'
PU,',',P.F=',PF:2:4,' ',STRIN:4);
                WRITELN('--------------------------------------------------------------------');
                NUMBER:=NUMBER+1;
             IF NUMBER=1 THEN
             BEGIN
             WRITELN;
             WRITELN('Find Capacitor(bus 2) with V1=V2.');
             WRITELN;
             WRITELN('Please enter Q2DC then angle X2(RAD) for approximation');
             WRITELN('Enter Q2DC value of between 14.33 to -5.7 for convergence');
             READLN(Q2DC,X2);
             WRITELN('  Q2DC = ',Q2DC:4:5,'   X2(RAD) = ',X2:4:5);
             WRITELN('COUNT':10,'Q2DC':8,'X2(RAD)':17,'V2C':14,'Pg':14);
             CP2:=1.0; Q1I:=1; Q1DS:=Q1I;  V22:=1;X22:=1;
             CQ2:=1.0;COUNT:=0; QGC:=5.0;PGC:=5.0;V2:=1;
             WHILE (ABS(CP2)>T) OR (ABS(CQ2)>T) DO
             BEGIN
             {*Q2DC:=-Y[2,2]*SQR(V2C)*SIN(PHI[2,2])-(Y[2,1]*SIN(PHI[2,1]+X1-
X2)*V2C*V1)*}
             a:=-Y[2,2]*SIN(PHI[2,2]);
             b:=-Y[2,1]*SIN(PHI[2,1]+X1-X2)*V1;
             c:=-Q2DC;
             V2C:=(-b+sqrt(sqr(b)-(4*a*c)))/(2*a);
             { IF V2C>0 THEN
             V2C:=V2C
             ELSE
             V2C:=(-b-sqrt(sqr(b)-(4*a*c)))/(2*a);}
             {IF V2C1}
             P2DC:=Y[2,2]*SQR(V2C)*COS(PHI[2,2])+(Y[2,1]*COS(PHI[2,1]+X1-
X2)*V2C*V1);
                CP2:=-P2DS-P2DC;
                {CQ2:=-Q2DS-Q2DC;}
                V22:=V2-V2C;
```

```
        J11:=Y[2,1]*V1*V2C*SIN(PHI[2,1]+X1-X2);
        J12:=Y[2,1]*COS(PHI[2,1]+X1-X2)*V1+Y[2,2]*2*V2C*COS(PHI[2,2]);
        J21:=Y[2,1]*V1*V2C*COS(PHI[2,1]+X1-X2);
        J22:=-Y[2,1]*V1*SIN(PHI[2,1]+X1-X2)-Y[2,2]*2*V2C*SIN(PHI[2,2]);
        J1:=1/J11;
        J2:=-J12/J11;
        J3:=J21/J11;
        J4:=J22-((J21*J12)/J11);
        X22:=(J1*CP2+J2*V22);
        CQ2:=(J3*CP2+J4*V22);
        X2:=X2+X22;
        Q2DC:=Q2DC+CQ2;  {Q2DC:=-Q2DS-CQ2;}
PGC:=SQR(V1)*Y[1,1]*COS(PHI[1,1])+V1*V2C*Y[1,2]*COS(PHI[1,2]+X2-X1);
QGT:=-SQR(V1)*Y[1,1]*SIN(PHI[1,1])-V1*V2C*Y[1,2]*SIN(PHI[1,2]+X2-X1);
        QGC:=QGT-(-Q1DS); {Pt=Pg-Pload)}
        COUNT:=COUNT+1;
        WRITELN(COUNT:6,Q2DC:15:5,X2:14:5,V2C:14:5,PGC:14:5);
        END;
        X2:=X2*180/PI;
        PL:=PGC-(-P2DC)-P1DS;
        QL:=QGC-(-Q2DC)-Q1DS;
        PF:=PGC/SQRT(SQR(PGC)+SQR(QGC));
        AG:=arctan(QGC/PGC)*180/PI;
        WRITELN('AG= ',AG:2:4);
        IF AG>0.00 THEN
        STRIN:='LAG'
        ELSE
        STRIN:='LEAD';
        Q2C:=-Q2DC-Q2DS;
        WRITELN(' THE ANSWER AFTER ',COUNT,' ITERATION IS : ');
WRITELN('BUS NO  VOLT(PU) ANGLE(DEG)  PD(PU)   QD(PU)   PG(PU)
QG(PU)');
        WRITELN('------ ------- ---------- ------ ------ ------ ------');
        WRITELN('1',V1:13:4,X1:12:4,P1DS:10:4,Q1DS:9:4,PGC:9:4,QGC:9:4);
        WRITELN('2',V2:13:4,X2:12:4,-P2DC:10:4,-Q2DC:9:4,'0.0000':9,'0.0000':9);
        WRITELN('PLOSSES=',PL:2:4,'        PU,                 ','QLOSSES=',QL:4:4,'
PU,','P.F=',PF:2:4,' ',STRIN:4);
        WRITELN('-------------------------------------------------------------------');
WRITELN('THE RATING OF THE CAPACITOR IS -Q2DC-Q2DS = ',Q2C:4:4);
        WRITELN;
      END;
      NUMBER:=NUMBER+1;
      END;
      END.
```

Chapter 8
Process Control

8.1 Water Tank Control System

The experiment we conduct studies the control response of the water level control
process in the system. We control the water level in the tank for the process control
system. Conventional PID control characteristics demonstrate the dynamics of the
system process. In most parts of the experiment, we only make use of one tank. So
we derived a first order equation G(s) for the main process. The apparatus as shown
in the diagram consists of the double tank system. A separation valve (A) locates
between the tanks. We fix the level and flow sensors on each of the tanks. We
supply a voltage input from zero to 10 V to each different tank. The supply voltage
is proportional to the water level and flow rate of each different tank. A plotter
connects to plot the control characteristics of the system (Figs. 8.1, 8.2, 8.3).

The error signal shows the differences between the set-point and the output. We
first study the proportional control system. By adding a PI controller into the
closed-loop system, we can derive the steady-state level output from the block
diagram with a step input.

8.1.1 First-Order Derivation

$$Q_0 = H/R \tag{8.1}$$

$$C\frac{dh}{dt} = Q_i - Q_0 \tag{8.2}$$

© Springer Science+Business Media Singapore 2016
T.S. Ng, *Real Time Control Engineering*,
Studies in Systems, Decision and Control 65,
DOI 10.1007/978-981-10-1509-0_8

Fig. 8.1 Single tank diagram

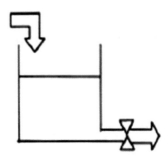

Fig. 8.2 System block diagram

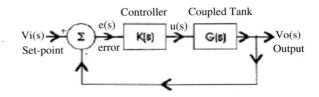

By substitution:

$$Q_i = C\frac{dh}{dt} + \frac{H}{R} \qquad (8.3)$$

Therefore:

$$Q_0/Q_i = \left(\frac{H}{R}\right) \Big/ \left(C\frac{dh}{dt} + \frac{H}{R}\right) = 1/(Ts+1) \qquad (8.4)$$

By using the controller Kp alone, the closed-loop transfer function is:

$$C.L\,T.F = Kp/(Ts+1+Kp) \qquad (8.5)$$

The steady-state output level due to a unit step input is (Fig. 8.5):

When Kp = 3: 3/(Ts + 4) = 0.75 V(plot showing 0.85 V)
When Kp = 6: 6/(Ts + 7) = 0.85 V(plot showing 1 V)
When Kp = 9: 9/(Ts + 10) = 0.9 V(plot showing 1 V)

By using PI controller:

$$C.LT.F = (Kp + Ki/s)/[Ts+1+(Kp+Ki/s)] \qquad (8.6)$$

When limit s → 0, Transfer function equals Ki/Ki.
The steady-state output level due to unit step input is reduced to 1 V.

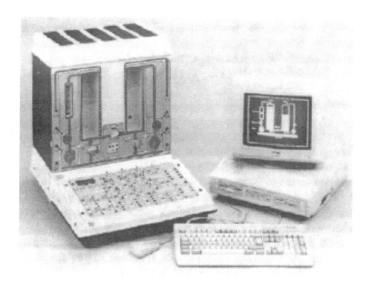

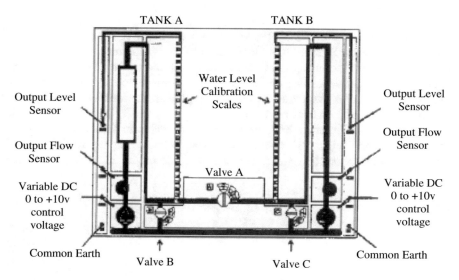

Fig. 8.3 Water tank apparatus

If Kp is fixed at 5 V (Fig. 8.6):

$$
\left.
\begin{array}{lll}
\text{When Ki} = 0.05: & 0.05/0.05 = 1\,\text{V} \\
\text{When Ki} = 1: & 1/1 = 1\,\text{V} \\
\text{When Ki} = 5: & 5/5 = 1\,\text{V}
\end{array}
\right\} \quad \text{plot is 1 V}
$$

If Ki is fixed at 0.5 V (Fig. 8.7):

$$\left.\begin{array}{lll}\text{When Kp} = 2.5: & 0.5/0.5 = 1 \text{ V} \\ \text{When Kp} = 5: & 0.5/0.5 = 1 \text{ V} \\ \text{When Kp} = 7.5: & 0.5/0.5 = 1 \text{ V}\end{array}\right\} \quad \text{plot is 1 V}$$

8.2 Single Tank Control

The actual response due to a step input is as shown in the plots. The time constant we plotted in Fig. 8.4 shows 1.75 min (105 s). For the proportional controller alone, the steady-state error reduces as Kp increases. We manually offset the input, to have a clear readout for the differences in the steady-state output plotted. The actual plot might have slight variation from the derivations for the output level. It is partly due to the pump's overhead loss or the plotter pen mechanism loose. Furthermore, small particles present in the liquid might block the pump suction head. For a small unit input of 1 V, all these might be significant to cause the output deviation. The disadvantage of using the proportional Kp term alone is that it introduces an offset in the final value as shown in the Fig. 8.5. So we introduce the proportional-integral PI term. It is clearly seen by derivation and from the experiment plot also, that the steady-state error is being reduced to zero. The PI controller is properly selected depending on the required response by choosing the Kp term first then the Ki term. As shown in the graphs plotted, we found that the system exhibits second-order characteristics from the PI controller. With a fixed Kp of 5, the overshoot gets larger as Ki increases. We are unable to fix the Ki first then adjust the Kp term. Figure 8.7 depicts the reasons behind. Ringing oscillations are introduced into the system when Kp is small. Moreover, there is always an

Fig. 8.4 RC time constant

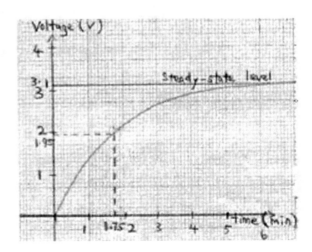

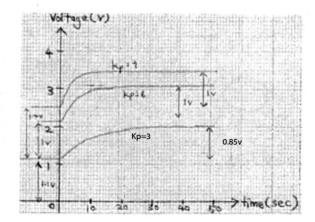

Fig. 8.5 Proportional alone

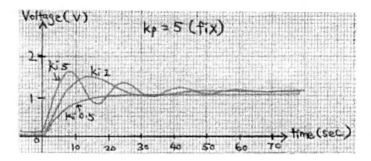

Fig. 8.6 PI output (fixed proportional)

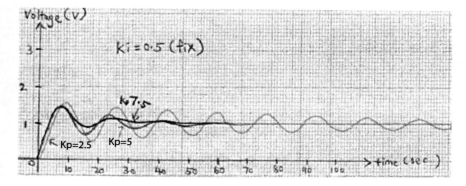

Fig. 8.7 PI output (fixed integral)

overshoot present in the system, which we are unable to eliminate. Thus, Fig. 8.6 shows the correct tuning method to adjust the steady-state final output level without exhibiting any overshoot. From the PI controller, we can add on an additional derivative term into the system. The PID controller can be introduced to improve the system characteristics further. By adding the derivative term, it increases the overshoot, at the same time reduces oscillations and ringings. Moreover, it also reduces the time constant at the transient stage. Therefore, it improves the speed of responses to reach steady-state. So, we learned that the derivative term is used to compromise the integral term in the PID controller. It enables three selectable tunings in the PID controller, for a more precise control of the system.

Chapter 9
Machine Learning

Global competition, changing customer's needs, increasing product reliability and complexity, volatile economic conditions and higher customer expectations are the changing set of business requirements that causes the need to optimize productivity, improving quality and reducing cost. The importance of neural network implemented is to boost the process methodology. We monitor the process parameters in the record flow chart plotted to analyse and study the progress. Next, suitable controllers can be implemented. Automatic corrections will be feedback to the system for the final desired output. Likewise, the process visualization for the plant can be simulated by software in real time.

9.1 Neural Network in Process Control

A neural network learns your process by observation and adapts its strategy for a more precise control. Neural networks are self-tuning systems that automatically take advantages of process upgrades and compensate for system degradation. Neurocontrol comes into existence to the problem solving of tuning a noisy, non-linear and complex system. Also, it can reduce the cost of implementing solutions for the problem. The nonlinear and multivariable capabilities of neural networks make the technology ideal for direct process control [24] (Fig. 9.1).

The artificial neural network has successfully used in many process control applications. It allows complexity control and critical monitoring of the process plant and sensors. In many systems, performance degrades over time due to deterioration of the system components. For compensation, operational parameters are dynamically tuned to optimize system performance. An ANN can be used to make decisions about the system operation and adjust the appropriate control to keep the process operates with optimal efficiency. An advantage of ANN over the traditional adaptive controllers is that we can continuously update the ANN with new

© Springer Science+Business Media Singapore 2016
T.S. Ng, *Real Time Control Engineering*,
Studies in Systems, Decision and Control 65,
DOI 10.1007/978-981-10-1509-0_9

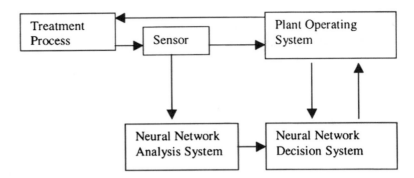

Fig. 9.1 Neural network process control

information by using a dynamic learning approach. The backpropagation algorithm is commonly used to train the ANN with the training data which composed of historical data about the process.

9.2 The Artificial Neurons

The neural network comprises of artificial neurons which are group into three main layers: input, output and hidden layers. Its function is by the rules of mathematical equations built into a semantic rule-based implementing as a controller to navigate the system and monitor its progress through its self-iterative learning process.

The information processing performed may be taken as the signals appearing at the unit's input. Or the action potentials to the synapses. The effect (PSP) of each signal can be approximated by multiplying the signal with some number or weight to indicate the strength of the synapse. The weighted signals are now summed to produce an overall unit activation. The unit will produce an output response when the activation exceeds a certain threshold. This functionality captured in the artificial neuron the Threshold Logic Unit (TLU) (Fig. 9.2).

$$A = W1X1 + W2X2 + W3X3 + \cdots$$
$$A = \sum_{i=1}^{n} WiXi \tag{9.1}$$

$$\left. \begin{array}{l} Y = 1 \text{ if } A \geq \theta \\ Y = 0 \text{ if } A < \theta \end{array} \right\} \tag{9.2}$$

We suppose there are n inputs with signals X1, X2, etc. The signals have a boolean output that is '0' or '1'. The threshold value is assumed zero. We illustrate the g value in Fig. 9.3.

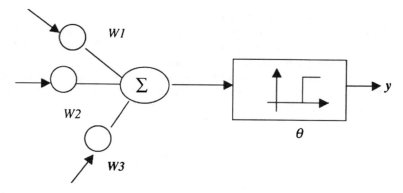

Fig. 9.2 Neural network

Fig. 9.3 Activation graph

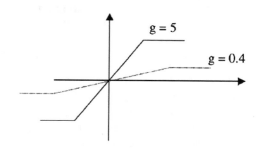

The graph shows the difference between the two g values. The g value is higher when the slope is steeper. For calculation, the g value is always put to one. The A value is always negative.

Thus the equation:

$$f(x) = \frac{1}{1 + e^{\wedge(g*(A-\theta))}}$$
$$= \frac{1}{1 + e^{\wedge A}} \quad (\text{Where } \theta = 0 \text{ and } g = 1) \tag{9.3}$$

9.3   Techniques Involved in the Controllers

Generally, there are five basic approaches in neuro-control techniques. They are:

(1) Supervised control
(2) Direct inverse control
(3) Neural MRAC-type adaptive control
(4) Reinforcement learning
(5) Unsupervised control.

(1) Supervised control: We make use of a training set consisting of X(t), and u*(t) where u* is the target action vectors. The system records the actions as well as the sensor inputs so that, we can track the responses. We need to calculate the deviation error for the weight adaptation.

(2) Direct inverse control: It is according to supervised learning. The states of the system make up the input of the plant, and the targets are the actuating signals. It is suitable for robot control.

$$X1, X2 = f\ (Q1, Q2)$$
$$Q1, Q2 = f^{-1}(X1, X2)$$
(9.4)

(3) Neural model reference adaptive control (MRAC): A reference model is supplied to output the desired trajectory. The system is linear, and the parameters of the plant are not given or predicted. The technique is near to direct inverse control. Specified equation:

$$U = -0.5 \sum_{t,i} (Xi(t) - Xi * (t))^{\wedge}2$$
(9.5)

(4) Reinforcement learning: The output feedback of the plant either aids or negates the control states depending on the environment reaction. The aiding signal reinforces those states that contribute to improvement, while the deducting signal, reduces the states that produce the improper behaviour. It uses a 'fuzzy' approach to the calculated output. The weight changing stops when the output stabilises in the range of the 'fuzzy' categorized output.

(5) Unsupervised control: The output of the subsets are put together at random or unsupervised, but the members of the same subset are one of its kinds. Thus, it is suitable for feature mappings based on the similarity of the input patterns.

From all of the above techniques, supervised control is used for neural network process control system as it enables on-line tuning, changing and process optimization in the system.

9.4   NN Learning Rules

(1) Hebb learning: If the two neurons take the same state at the same time (both inactive or both active), the output of the weight connection between them increases (Fig. 9.4).

(2) The perceptron learning rule: We calculate the output error from the deviation between the desired and actual output. Then, it is used to calculate the changes in weight in the connectism neurons (Fig. 9.5).

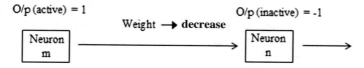

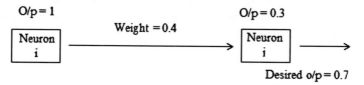

Fig. 9.4 Hebb learning

Fig. 9.5 Perceptron learning

$$\Delta \text{ weight} = \text{o/p i} * \text{deviation o/p}$$
$$\text{New weight} = \text{old weight} + \Delta \text{ weight} \tag{9.6}$$

(3)   The delta learning rule: It makes use of the same concept as the perceptron learning, except we divide the deviation errors by the number of the prepro-cess neurons (Fig. 9.6).

$$\text{deviation o/p} = \text{desired o/p} - \text{o/p}$$
$$\Delta \text{ weight} = \text{i/p neurons} * ((\text{deviation o/p})/(\text{number of i/p neurons}))$$
$$\text{new weight} = \text{old weight} + \Delta \text{ weight}$$

$$\tag{9.7}$$

(4)   Backpropagation learning rule: It uses the same tactic as the delta learning rule, but it allows the changing of weights in the additional hidden layers introduced. The derivative of the activation function is the sensitivity of the

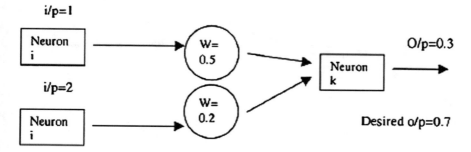

Fig. 9.6 Delta learning

activation function. For an output neuron, it is back propagated to the predecessor hidden neurons to sum up the interconnected weight error. The derivatives are used to calculate the actual weight changes for both the output and hidden neurons by multiplying the errors in it.

$$\text{Derivative of activation function:} \quad a' = a^*(1-a) \tag{9.8}$$

For all the above learning rules, backpropagation is most popular for process control as it enables a more precise manipulation such that we can change the weights in the hidden neurons.

9.5   Selection of the Learning Algorithms

There are two main types of neural nets:

(1) The feedforward neural net
(2) The feedback neural net

(1a) The perceptron net (feedforward net): Only the weights at the output of hidden neurons are trainable. This layer uses the perceptron linear threshold activation function. The specified equation is (Fig. 9.7).

$$\text{New weight} = \text{old weight} + (\text{o/p of predecessor neuron} * \text{target deviation} \\ * \text{learning rate})$$

$$\tag{9.9}$$

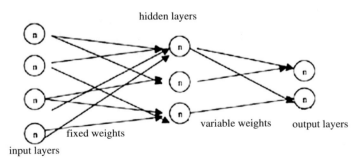

Fig. 9.7 Perceptron network

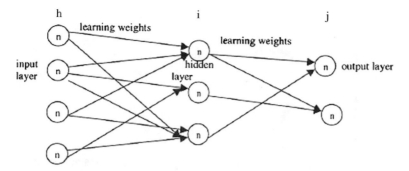

Fig. 9.8 Backpropagation network

The hidden layer in the system lies between the input and output layer by the interconnected neurons. There is a direction for the computation of information. The number of hidden neurons may be 50 % of the number of input neurons. It is not always the case to have the same number of input neurons as the output neurons in the last layer.

(1b) Backpropagation (feedforward net): It allows weights into and output of the hidden neurons to change during learning. There are sensings of activation neurons due to weight changes. There are two calculations involved. First, is to calculate the net output and the error in each neuron in the forward direction and secondly, to backpropagate the net errors to the preceding neurons. Besides, backpropagated errors to neurons in predecessor layer are held at the same time. Thus, the calculated net output formed is slow. Assuming, neuron i is the preceding of neuron j (Fig. 9.8).

Every unit of the neuron inherits with a non-linear function called the activation function or the standard sigmoidal function.

$$\text{Equation:} \quad f(\text{net}) = \frac{1}{1 + e^{\wedge -net}}$$

$$\text{where:} \quad \text{netj} = \sum_i wij\, oi \tag{9.10}$$

$$\& \quad f_i'(net) = o_i(1 - o_i)$$

We use the data training set to train the system for error reduction. At the same time, we use the validation set to validate the data. The optimal point, where the intersection occurs between the validation and training errors, is the point to stop training. This is the best generalization before the validation error starts to increase (Fig. 9.9).

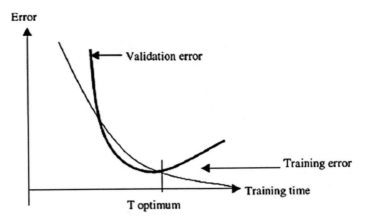

Fig. 9.9 Best generalisation graph [22]

We calculate the error function by the sum of squared error differences between the actual and targeted output value.

$$E = 0.5 \sum_{j=1}^{N} (t^j - y^j)^2 \tag{9.11}$$

$$\Delta w_i^j = \propto \sigma^j f_j'(\text{net}) \, (t^j - y^j) x_i \tag{9.12}$$

where $\propto$ represents the multiplier factor, σ is the learning rate, f(net) the activation function, x defines the input to the neuron. T is the target output and y is the actual output. We can assume the factor $\propto$ to be 1.

$$\delta^j = \sigma^j f_j'(\textit{net}) \, (t^j - y^j) \tag{9.13}$$

$$\Delta w_i^j = \propto \delta^j x_i \tag{9.14}$$

$$\Delta w_h^i = \propto \delta^i x_h \tag{9.15}$$

$$\delta^i = \sigma^i f_i'(\textit{net}) \sum_{j \in I_i} \delta^j w_i^j \tag{9.16}$$

where Ii is the set of nodes after the hidden node i. It is known as the fan-out of i.

(1c)   Radial Basis Function (feedforward net): The RBF have unique identity such that they have only one hidden layer with radial basis functions $\Psi(x)$ and produces only linear output. As such, training of the net consists of the unsupervised part to define the center of the radial basis function and supervised part to learn the weight. A disadvantage is that we often used it for the small number of inputs. However, for a large input multi-layer perceptron network, gives a better generalisation.

(2a) Hopfield Net (feedback net): The difference is that weight calculation leads to only the change of neuron states (pattern learning). The minimal problem should be modelled so optimizing the objective function minimised the energy function and that non-fulfillment of a constraint, leads to an energy increase. Its function is more suitable for pattern recognition. The equation shows the error function. Where Qi is the neuron input.

$$\text{Equation:} \quad -0.5 + \sum active\,neurons + \sum Qi \qquad (9.17)$$

(2b) Simulated Annealing (feedback net): Another name for it is the Boltzmann machine. It uses the same principal as Hopfield Net and is a further enhancement of it to counteract the disadvantages of this system. That is, it allows the neurons to change state to avoid local minimal of the system. It consists of visible and hidden neurons interconnected. This system stops at global minimum. However, the main drawback of the system is that it is slow and is used for travelling salesman problem.

9.6 The Network Topology

The neural network data processor consists of the input, hidden and output layers. The decision for the numbers of input neurons and the layers of the hidden neurons involved depends on the complexity of the system process. The data training set for the input neurons rely on how we control the system and the input sensors involved. The neuron numbers in the hidden layers and the number of hidden layers in the system plant will affect the system monitoring and accuracy predetermined from the neural processor. Thus, a careful determination of the number of neurons in the hidden layers are often chosen. The final part is the target monitoring of the plant. It consists of the last layer of the output neurons. The numbers of output neurons depend on how many and what kind of action output we are controlling. It is in term compared with the desired target value, and the deviation is feedback to the predecessor to re-calculate the output to be maintained at the target requirements. Therefore, a network topology is derived from a given system to be controlled.

9.7   MLP Backpropagation Network for Process Control

The neural network based controller and predictor are computerised systems, which perform optimization in the continuous chemical processes. It monitors the chemical products and adjusts the effect to produce the desired chemical reactions and properties in a plant. Most of the existing chemical plant do not have chemical sensors that can measure its properties. Human interventions are needed to examine the samples off-line and consistently adjust the plant operating parameters to suit the target. The neural based system can replace the human intervention to estimate the chemical properties of the final product on-line automatically. The SCADA system connected allows the display of the plant information. Human operators can then make decisions to take further action to counteract conditions such as degradation of the hardware sensors and tunings of plant parameters to maintain the desired target performance. An example is a chemical process in a mixer plant. We can use the system for forecasting as the output produced is the future value of the input signal (Figs. 9.10, 9.11, 9.12).

Based on backpropagation calculations the following data set were being tabulated for round one of the backpropagation learning:

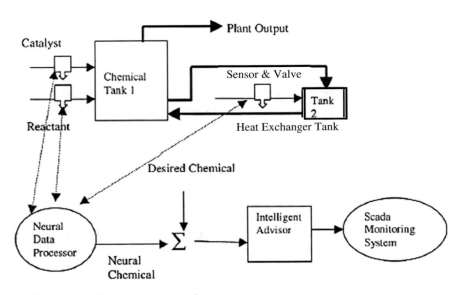

Fig. 9.10 Mixer plant neural net control system

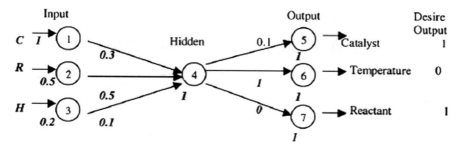

Fig. 9.11 Neural net system processor

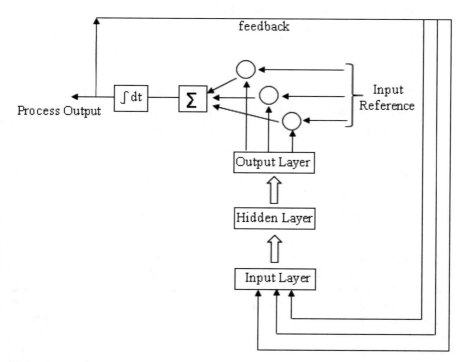

Fig. 9.12 Neural network external structure

$$\text{gin4} = 1*0.3 + 0.5*0.5 + 0.2*0.1 = 0.57$$

$$\text{act4} = \frac{1}{1+e^{-0.57}} = 0.639$$

$$\text{out4} = 1$$

$$\text{gin5} = 1*0.1 = 0.1$$

$$\text{act5} = \frac{1}{1+e^{-0.1}} = 0.525$$

$$\text{out5} = 1$$

$$\text{gin6} = 1*1 = 1$$

$$\text{act6} = 0.73$$

$$\text{out6} = 1$$

$$\text{gin7} = 1*0 = 0$$

$$\text{act7} = 0.5$$

$$\text{out7} = 1$$

$$\text{D5} = [0.525*(1 - 0.525)]*(1 - 1) = 0$$

$$\text{D6} = [0.73*(1 - 0.73)]*(0 - 1) = -0.197$$

$$\text{D7} = [0.5*(1 - 0.5)]*(1 - 1) = 0$$

$$\text{D4} = 0.639*(1 - 0.639)*[0.1*0 + 1*(-0.197) + (0*0) = -0.045$$

$$\Delta \text{W54} = 0*1 = 0$$

$$\Delta \text{W64} = -0.197*1 = -0.197$$

$$\Delta \text{W74} = 0*1 = 0$$

$$\Delta \text{W41} = -0.045*1 = -0.045$$

$$\Delta \text{W42} = -0.045*0.5 = -0.0225$$

$$\Delta \text{W43} = -0.045*0.2 = -0.009$$

$$\text{W54} = 0.1+0 = 0.1$$

$$\text{W64} = 1 - 0.197 = 0.803$$

$$\text{W74} = 0+0 = 0$$

$$\text{W41} = 0.3 - 0.045 = 0.255$$

$$\text{W42} = 0.5 - 0.0225 = 0.4775$$

$$\text{W43} = 0.1 - 0.009 = 0.091$$

The Table 9.1 weights tabulation is in accordance to Fig. 9.11. That is, it depends only on one hidden neural processor. Computation begins until stabilisation once the three adjustable parameters for the three valve sensors are feedback into the neural network input. The neural network shows the final values during stabilization at the seventh round (1 epoch meaning the NN completes 1 cycle for the total nos. of training examples). So the desired output is also achieved. We monitor the progress and outcome at the control station. The system shows the on-line tuning using backpropagation technique to match the desired output of the plant.

Table 9.1   Weights tabulation

	0	1	2	3	4	5	6	7
W41	0.3	0.2546	0.2144	0.1823	0.1614	0.1543	0.1622	0.1622
W42	0.5	0.4773	0.4572	0.44115	0.4307	0.42715	0.4311	0.4311
W43	0.1	0.09092	0.08288	0.07646	0.07228	0.07086	0.0724	0.0724
W54	0.1	0.1	0.1	0.1	0.1	0.1	0.1	0.1
W64	1	0.8034	0.5898	0.3603	0.1182	−0.1309	−0.3798	−0.3798
W74	0	0	0	0	0	0	0	0
D5	0	0	0	0	0	0	0	0
D6	0	−0.1966	−0.2136	−0.2295	−0.2421	−0.2491	−0.2489	0
D7	0	0	0	0	0	0	0	0
D4	0	−0.0454	−0.0402	−0.0321	−0.0209	−0.0071	0.0079	0
O4	1	1	1	1	1	1	1	1
O5	1	1	1	1	1	1	1	1
O6	1	1	1	1	1	1	0	0
O7	1	1	1	1	1	1	1	1
I1	1	1	1	1	1	1	1	1
I2	0.5	0.5	0.5	0.5	0.5	0.5	0.5	0.5
I3	0.2	0.2	0.2	0.2	0.2	0.2	0.2	0.2

W Weight; D Deviation; O Output; I Input
Continuous neurons are used for the input values
$I1$ Quality of catalyst being added to the reactant being encoded to the continuous value of 0–1
with respect to position of the valve
$I2$ Reactance input to the tank being encoded to the range of 0–1
$I3$ Range of heat measurement being encoded to the range of 0–1 (from 0 to 100° respectively)
Binary neurons are used for the target output
$O5$ Catalyst density is set to range of 1–5 units (binary 0)
Catalyst density is set to range of 6–10 units (binary 1)
Target at above 5 units (binary 1)
$O6$ Temperature is set to the range of 0–35° (binary 0)
Temperature is set to range of 36° and above (binary 1)
Target at below 36° (binary 0)
$O7$ Reactance quantity finalized to the range of 2–3 at binary 1

9.8   Chemical Plant NN Feedback Control System

The reaction states of the catalyst and the reactance were difficult to analyse. Instead, their properties and characteristics were collected from laboratory data analyst. Temperature input is sensed using the temperature sensor. Besides setting the chemical target quality, the catalyst output density and the reactance quantity are also set according to the control temperature. However, there arise the difficulty in determining the parameters of the non-linear system. So we employ the neural network system to control the variables for the catalyst, reactance and the temperature, to reach its desired states. The neural network can counteract the system error due to the hardware problem of the mechanical valves or sensors. The weights adapt automatically to the targeted set-point even if the input jams at a position due to failure. The signal from the neural network feeds into the SCADA (Supervisory Control And Data Acquisition) system and monitor accurately. The neural network can also control their output to their final states by using PID controller to adjust the final control element for the catalyst and reactance. We can include the network advisor to create alarm and leads to human intervention when the controlled outputs were absurd. The system thus acts as a safeguard to monitor the on-line process and estimates the final product with references to the desired requirements. It serves as a simulation tool also, to test different operating process conditions. It is particularly advantages when justifying process enhancements and modifications. The final result will be the product with the desired chemical properties. However, the speed, number of iterations, accuracy and convergence for the neural network depend on the careful design of the neurons used. The output of the desired state can also be in real neurons to control their exact value instead of binary neurons to control a certain range. System lag time contributes slightly to the drawback of the system. The delay time to activate the final control element from the neural network or vice versa when sensing creates an interval to real time reaction. However, in water process control, system lag time is not critical as we do not expect a fast response from liquid flow/level turbulence and stabilization.

9.8.1 Process Design

From the case above, we can design a feedback with a sampling period, for example, one sample per 20 s, to input the detected parameter values into the neural network system. At the same time, we can also set an integrated delay of 10 s to output from the NN controller to the plant. The delay allows for the training and adaptation of the neural network computation. The neural network only activates once it detects the set of input values. It then goes through the neural network layers to train for the desired output. We initialize the weights to random values, and neglect the thresholds in the NN system. Then we perform an off-line simulation to compute the desired preset NN values for the on-line system. That will reduce the online adaptation training time of the neural network system. The output of the neural network is sent to control the valves, after going through the delay. The system forms a closed loop control for each sampling cycle of the NN control system.

The Rosemount wireless flow, level and temperature transmitters [28] are deployed at the field site to do the job. The control station will receive the wireless transmission signals for the three parameters feedbacked. The closed loop system performs sampling at 0.05 Hz for the neural network computation. After going through a delay sampler, the three output valve parameters are feedback instantly into the field controller through wireless transmission detector. Each of the detected signals goes through a signal conditioner circuit to scale the signal suitable for the actuating element. We can also place the signal converter at the NN station before transmission. The output of the scaled signal is fed into the PID controller to adjust the control valve in real time. After twenty seconds, the flow/level/heat of the reactance and catalyst will be feedbacked into the neural system wirelessly. Before that, we can calibrate each of the flow/level/temperature transmitters with their valves' positionings. Once calibrated, every valves' positions will correspond to their respected variables' values. The NN system compares these quantities with the desired amount. The valves' control system and the neural network controller can say to operate simultaneously (depending on the selection of the system sampling frequency) during the continuous closed loop cycles. The speed of the feedback control system is set faster by the sampling period of the system. In doing so, we also had to reduce the delay feedback output of the NN.

9.8.2 Process Verification

We collected eight input/output sets of random samples for the 3-1-3 neural network configuration. These samples were collected randomly from the different values to the input neurons' layer [1 0.5 0.2;0.5 0.5 0.5;0.5 0.5 0.2;0.5 0.5 1;0.5 0.37 0.5;0.5 0.59 0.5;0.33 0.5 0.5;1 0.5 0.5] for matching to the same desired output [1 0 1]. With the desired output fixed, the neural network backpropagated itself. Each set of its output parameters stabilized itself automatically after adaptation. We record the weights for each different sets of the input/output. The output neuron parameters were plotted in the graph, through the formula as follow:

$$\text{Catalyst} = W54\left(\sum_{i=1}^{3}\{W4i\,Xi\}\right) \tag{9.18}$$

$$\text{Temperature} = W64\left(\sum_{i=1}^{3}\{W4i\,Xi\}\right) \tag{9.19}$$

$$\text{Reactance} = W74\left(\sum_{i=1}^{3}\{W4i\,Xi\}\right) \tag{9.20}$$

Note: W4i refers to the neurons' connected weighing values between the input layer and hidden layer; X1 is the catalyst input; X2 the reactance input; X3 defines to the temperature input signal (where Xi references to each of its valve's variable).

The catalyst output value may vary during tuning, to match the desired reference output of the neural network. That goes along with the auto adjustment of the temperature for the reaction. As can be seen in Fig. 9.13 from sample 7 to 8, the temperature automatically reduces as we increase the catalyst input. That is auto tuning of the system to maintain the chemical reaction of the mixture we set. The two parameters go according to a scaling rather than its unit. As we know, the

Fig. 9.13 NN output neurons (with secondary axis)

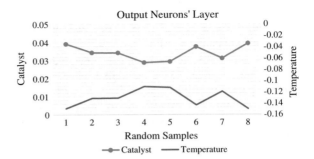

function of the catalyst is to accelerate the reaction of the chemical mixture. So, more catalyst will lower the activation energy require for more chemical reaction to occur. Less catalyst makes the reaction slower. Thus, we increase the temperature of the reaction or mixture to boost back the chemical reaction in the plant, to produce the desired chemical mixture. The amount of heat boosted is equivalent to the catalyst reduction during valves' auto tunings. With a static valve positioning for the reactance, the catalyst valve and the coolant valve auto adjusts themselves to balance the mixture reaction required in the plant. The valves' adjustment affects the rate of the reaction and the quality of the mixtures, within the safe range of the chemical production. Our aims are to meet these two requirements in the system. Alternatively, our reference output for the neural network is provided to match these two criterions. An advantage of the neural network control system is that we do not need to calculate the error relationship between the valves' settings and their corresponded quantities of its product parameters. The continuous auto tunings of the NN and the valves will reduce and eliminate the error deviation of the plant parameters. Moreover, the control station will calculate the matching between the actual plant and the desired chemical to correct the target signals. Figure 9.14 shows both the catalyst (top) and the temperature (below) for a set of the target. The NN output neurons are calculated using Eqs. (9.18)–(9.20), from the eight different input samples, to produce the same output target. The centerline shows the

Fig. 9.14 Neural network output layer

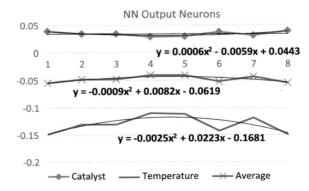

calculated average of the two parameters. We estimate the three curves by polynomial equations. With the given average and any of the parameter in any timeline, we can calculate the other parameters. The parameter found will lie along the polynomial of the parameter itself. In this case, we can calculate either of the parameters offline, to set the predictor limit for the parameters. Thus, we can predict the parameters' timelines also.

9.8.3   Process Improvement

So far, we have used the binary neurons for the NN system. Each neuron's output value is adjusted away from its real value during computations, for the neurons' output function. Hence, the system accuracy is affected. Previously, we have set the output layer neurons to its binary estimation. We lost the precision of the target setting also, due to the estimation to the target binary. We want to improve the system precision and accuracy. So we will now set the target output to their exact real values, with a precision of two decimal points. For each neuron, we will use their activation (sigmoid) function values as its output. Besides, we also update the threshold values using the update formula as shown.

$$\theta^j(t+1) = \theta^j(t) + \Delta\theta \tag{9.21}$$

$$\Delta\theta = \alpha(\theta^j(t) - \theta^j(t-1)) + \eta\delta^j \tag{9.22}$$

where η is the gain term; α is the momentum term; θ is the thresholds term of each node.

[Recall: Eq. (9.13) where $f_j(net) = \frac{1}{1+\exp(-(netj+\theta^j))}$]

For example, we set our desired target quality as [0.93 0.35 0.71]. Upon initialization, the starting weights are similar to Table 9.1. The initial threshold values biased at 0. We simulate for the first eight sampling inputs as in the program 9.1. We train the system and record their weights and biased values, after meeting its performance goal. Our mean square error set at exponential −24. According to the Eqs. (9.18)–(9.20), we trace the parameters' updated trends according to each of its input samples. We verify the first three input sampling sets [0.15 0.47 0.17], [0.42 0.23 0.74], [0.41 0.58 0.2] for the trend. For the first to the second input sample, we can see the rise in the catalyst but fall in the reactance. Besides, the temperature rises for the input samples. As the transitions for the input samples are the valves' parameters which take place at the site, it enables more of the chemical reaction to occur at the site. The neural network system automatic adjust itself to control the mixture to meet the target quality set. Thereby, causing the catalyst and the reactance mixture to reduce together with the targeted temperature. As we can see in Fig. 9.15, the catalyst drops in accordance to the reactance. So the mixture desired outcome is maintained. The target temperature for the mixture also

Fig. 9.15 Auto adjustment trends of the NN system

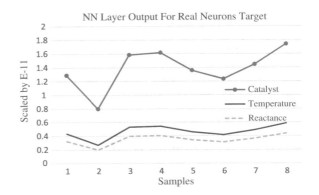

maintains throughout the mixing. For the 2nd to 3rd sample input, the reactance increases and the temperature dropped. It results in the slowing down of the chemical mixture. The resultant NN control reacts to boost back the chemical reaction to match the targeted parameters and chemical quality. We can clearly observe the rise in the parameters as in the figure.

9.9 Remote Operated Neural Network Control Plant

9.9.1 Field Instrumentations

The Remote Operations Controller (ROC800) [25] is ideal for applications in remote monitoring and control measurements. The 32-bit microprocessor-based controller is used widely, especially for PID flow control applications. The recorded database allows for the storing of events and the triggering of alarm signals in the plant. We can measure up to a maximum of 12 monitoring signals for the instrument. The wireless network radio modem provides the means for transmitting information for the ROC800. The features of the system allow it to operate with an update rate from 1 s to 60 min. The range of operating distance is from 230 m to 1 km. With the attractive characteristics, we can configure the ROCLINK 800 software to monitor the plant parameters. We can customize to control the plant also with neural network control capability (Fig. 9.16).

The ControlWave Micro is with the IEC 62591 module installed. Software in the control station manipulates the received signals for transmission back to the field. OpenBSI software can configure the ControlWave Micro. There are several choices of radio transmissions for the field parameters. The ROC mentioned is for discrete signals only. Besides, we can use the Rosemount 248 wireless temperature transmitter and the Fisher 4320 wireless valve position monitor for the system. Additional choice includes the wireless smart THUM adapter [27] to sense the valves' parameters. These devices are suitable for use with the IEC 62591

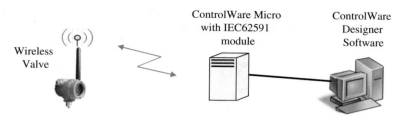

Fig. 9.16 IEC 62591 interfacing network

(WirelessHART). However, network radio module (NRM) [26] provides bi-directional transmission of measured and controlled signals. It allows for monitoring and intervention within the field operations. We can connect up the 2.4 GHz NRM for the purpose with an update rate of 0.05 Hz. The valve scaled position feedback calculates as:

$$\text{Scaled Position} = [(\text{Upper Range Value} - \text{Lower Range Value}) \\ * \text{Position} \% + \text{Lower Range Value}] \tag{9.23}$$

The valve sensor can also provide valve on/off control for overflow prevention in the plant system. Moreover, it can be used to open/close the coolant valve for the heat exchanger when the temperature reaches the target limit. The Westlock valve ranges a distance of 110° positioning turns during on/off. The linear or rotary valve reports every 1.5° turns. However, the transmission range is limited to only 100 m with a router.

9.9.2   Scaling and Conversions

The scaled value Eq. (9.23) from the valve will go through a converter unit to allocate the range into suitable values for the input neurons. Before that, the scaled value had to be in units of angular degrees. The scaled value is equivalent to the valve scaled position. The neuron range of 0–0.73 derives from the valve operating range of 0–110° turns. The converted value lies within the range of 0–1 for the neuron inputs.

$$\text{Converter value} = ([\text{Scaled Value} - \text{Lower range value}]/1.5) \times 0.01 \tag{9.24}$$

So the valve sensors' signals go through a conversion for matching the NN input range. For the output layer neurons' signals, we had to convert back the values to the scaled values using Eq. (9.24), if the valves do not read the actual neurons' values. We can refine the system to produce a more sensitive output. To do that, we just change the output functions of the neurons to real values instead of binary values. We can use the 'logsig' logarithm sigmoid function so that the output range stays between '0' and '1'. All these can be done at the control station for the input and output of the neural network system before transmissions.

9.9.3 Control Valves

The field electronic valve automation may consist of a single valve automation unit or with another valve adapter for transmitting readings back to the control station. A controller situates inside the electronic valve (see Fig. 9.18). It activates the PID input, which receives the signal through an antenna. The actuators then manipulate the valves from the PID signals. Altogether, at least, two throttle valve automation sets are required for the system. Either the on/off valve or the throttle valve are used for the heat exchanger to control the temperature of the mixture. Additional automatic wireless on/off control valve can be used for the system also. It is for the purpose of controlling the upper and lower limit of the chemical tank. The automated on/off valves for the mixer tank control the liquid overflow. We can use the digital set-point to trigger the overflow alarm as well as to control the automated on/off valve. A 4–20 mA level transmitter can be used to detect the level in the tank. We can align the signal and scaled it to the 0–0.73 range for the digital valve. Such that when the level is low at 0 mA, the valve will receive a 0.73 digital signal to open the valve fully. If the level reaches the full 20 mA signal, it will activate the valve with a zero signal to close it (Fig. 9.17).

9.9.4 Wireless Transmissions

The NN control system can be programmed to perform the function. We can insert a delay sampler between the input neurons and the input weights of the neural network. We program the system to detect the input signals from the valves at every 20 s interval. It is the same as delaying the system for every 20 s. Another way is to

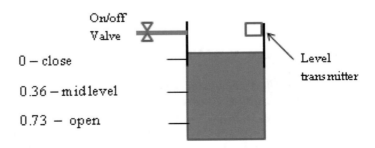

Fig. 9.17 On/off valve control

Fig. 9.18 Wireless valve automation [32]. Courtesy of Emerson Process Management

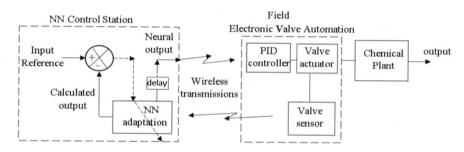

Fig. 9.19 Automatic neural network control system

set the sampling period for the wireless transmitting rate to 0.05 Hz. The NN will collect the input values to tune it to its desired output. The zero deviations in the output signals halt the NN from further training. Then, each output parameters are transmitted wirelessly to the valves. We set a transmitting delay at the output of the NN system to the valves. This delay can be set to 10 s or so, as long as it lies within the system sampling period (set at 20 s). The delay time allows for the automated valves' adjustment turns, as well as the valve sensor feedback and wireless transmitting times. Moreover, it also caters for the neural network training and adaptation time.

The output sampler turns on again, and the valves accept new input values every 20 s. The closed loop system controls the valves for the desired mixture chemical with controlled temperature limit (Fig. 9.19).

German engineering 'JULABO' [37], offers the state-of-the-art temperature control technology for a wide range of temperature measurements from −95 °C to over 200 °C. Several challenging features suitable for lab-based and integrated outdoor environment, especially for the chemical reaction. Features such as fast heat-up and cool down times for low power electronic controlled smart pumps provides a more reliable control technology for heat exchange units. Moreover, the advanced control system operates without changing the bath fluids. Wireless temperature control is available for the chemical process control system. We can easily integrate the system for process optimization.

9.10   Valves and Chemical Plant Tunings

9.10.1   Desired Chemical Mixture, Samples and NN Data

The reference input to the system is the desired output of the neural network. We conduct laboratory experiment for the quantity of the reactance [34] and catalyst mixture allowed. The mixture is tested to match the quality requirement. We record the required percentages of the mixture to produce the least heat temperature. A proportional amount of the mixture is carried out in the actual plant with the temperature monitored. We then collect the desired sample of the catalyst, reactance and heat requirements without the NN system involved. Besides, we also collect several input/output desired sets of the three valves' settings as input parameters for training the neural network. The valves' settings and its parameters are the same. Their ranges are between 0 and 1. The NN system is trained off-line to tune the adapted weights and biased parameters to their stable sets of values. Finally, we are ready to apply these NN parameters to the on-line system. In this way, the online training time can be reduced.

9.10.2   Chemical and Valves Calibration

The control station sets the matching and desire three valves' positionings with their parameters (catalyst, reactance and temperature) for the chemical mixture. The flow and level transmitters of the two small tanks transmit their signals back to the station and compare the quality of the mixing chemical with the proportional chemical property, collected from the lab samples. Besides, the temperature signal at the mixture tank also transmits back to the control station. The quality relates to the quantities of the reactance and catalyst mixed at the desired temperature. We first calibrate the temperature of the mixer tank with the heat exchange control valve, with reference to the catalyst and reactance. The flow and level of each tank determine the quantities of its chemical. Once the quality of the mixture matches at the desired temperature, the ideal valves positionings for the three control parameters will be set. Calibration can be at the control station, or at the field valve automated system or both. So we control the amount of catalyst, reactance and heat produced in the chemical mixing plant. The neural network system controls the three control valves as shown in Fig. 9.20.

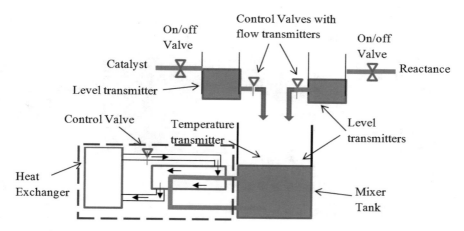

Fig. 9.20 Chemical plant valves and sensors

9.10.3   Trial Test in Actual Plant

The offline calculated NN parameters deploy to the real-time system. The trial test is conducted at the actual plant to get the correct output that we want for the complete system. We can adjust any initial settings of the three valves' parameters for the start. For every 20 s sampling delay, the NN receives three input values from the valve sensors to calculate at least an iteration in the NN. Through the NN training, the system can compute the correct output at the output neurons. These outputs transmit through the delay, and then to the site control valves wirelessly. The valves are adjusted automatically to produce the desired mixture within the controlled temperature range. We verify the actual property of the mixture at the plant output. The trial test completes once the quality satisfies.

There might be a problem that we are unable to boost the temperature up or cool it down for the mixer tank. For example, when the catalyst input drops, the heat might not be raised high enough to boost back the chemical reaction to take place. Remember that we only use a coolant heat exchanger unit in the plant. Hence, our remedy is to reinforce another heat exchanger to the system. The second exchanger unit is to supply the heat source to raise the temperature enough for the chemical reactions. So this heat exchanger is sourced with hot water instead of cooling water. The control command for the valve of the second heat exchange unit (heat source) will be opposite as the first (coolant source) unit. We define the valve manipulation as:

$$C_{CE} = 1 - C_{HE} \tag{9.25}$$

where C_{CE} is the coolant source, and C_{HE} is the heat source, they both control the valve position variables of the heat exchanger units. Furthermore, we may use four

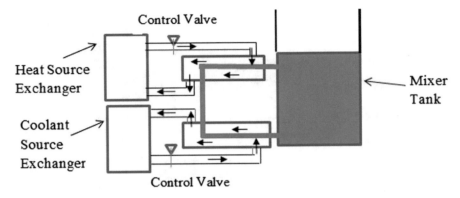

Fig. 9.21 Two heat exchanger units

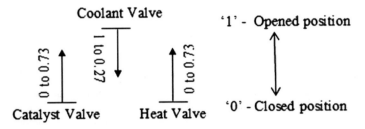

Fig. 9.22 Control valves operating ranges

input/output neuron signals for the neural network system to control the chemical plant. The additional input/output neuron variable references to the hot water exchanger control valve. We can then fine tune the internal neuron layers if it is better than the 4-1-4 NN configuration we proposed. The control valves operating ranges for the reactance is set similarly as the catalyst valve as shown in Fig. 9.22. Whereas, we set the two heat exchanger valves in the opposite direction as according to the definition Eq. (9.25). The cooling valve will not set to fully close while the heating valve will not be set to fully open. These settings can lower the presence of heat in the reactance at the initial stage (Fig. 9.21).

9.11 Computerized Neural Network Control System

9.11.1 NN Real Time Control Plant

Finally, we can implement the neural network to the real-time control system. Training and adaptation are necessary on the online system to eliminate any transmission network, instrumentations software and sensors errors. We can fit in the found NN parameters previously to reduce the training time. The NN parameters are automatically trained online again to adapt to the desired output of the three parameters. The NN outputs are found and transmitted to tune each of the corresponding valves. The continuous wireless closed loop network calibrates itself to maintain the valves' positions for the desired chemical output. In the system, we can be aware of the on-site valves or instrumentation hardware failures if we detect any changes in deviation in the actual chemical output with the set target (i.e. we check for the matchings between the target quality (set valves) and the actual chemical output quality, from the computerized advisor). Another advantage is that we can change the desired output online by changing the reference signals to the output neurons. The system will calibrate itself to its new weights and biased values. Thus, the control valves changed according to the new values.

9.11.2 Neural Network Control Valves

For the program, we input the chosen catalyst value for the plant. So the selected catalyst valve position is referenced or proportional to the catalyst variable. Unlike the catalyst and the reactance, the temperature is inversely proportional to the heat exchange valve position. Thus, we inverted and scaled the temperature output neuron value to control its valve position. We can control the heat exchange valve to perform the same way as the catalyst valve position variable. First, we need to calibrate the temperature according to the valve's variable. We calibrate the temperature of the desired chemical tank output, with reference to the heat exchange valve position. For example, a valve range from 0 to 0.73 is calibrated to the temperature from 0 to 73 °C. We can adjust the degree of warm water in the heat exchanger unit. Scaling is not necessary if the neurons' value matches the valves' range. The converted valve's variable ranges from 0 to 0.73, for the heat valve, disregards to the actual temperature readout in the mixer tank. The computer station does the conversions. For the coolant valve, we convert the variable for the valve positioning, by using the formula Eq. (9.24), with reference to Fig. 9.22. The heat exchange valve position is transmitted back to the input neuron with the same conversion and scaling as at the temperature output neuron. We maintain the reactance input set valve position at a level of percentage tolerance (at $\pm$ desired valve setting). Similarly, as the catalyst parameter, if we convert the output neurons of the reactance to the control valve, it has to be inverted back from the control valve variables to the input receiving neuron. Remember that we set and convert all the three parameters of the valve's positioning values to the input neurons. The neurons accept values from 0 to 1 for computation. Figure 9.24 shows the training convergence for the 3-1-3 neural network configuration. After training, we simulate the result to a random set of input for the three neurons. The simulation converges

Fig. 9.23 Neural network
I/O block

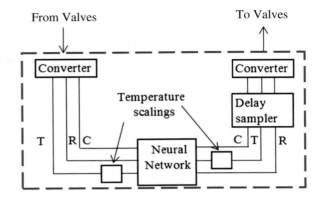

Fig. 9.24 NN training result

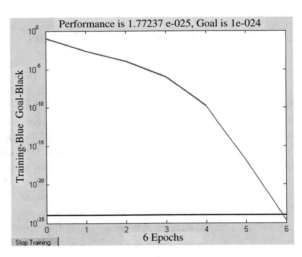

Fig. 9.25 Simulated result

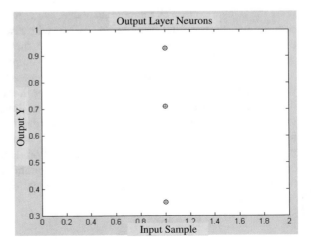

to the target output after six epochs. We record the training weights as shown. The output Y matches our desired values [0.93 0.35 0.71]. Figure 9.25 illustrates the output result. However, we can increase the numbers of target tracked at the same time by increasing the hidden neurons in the program (Fig. 9.23).

Program 9.1: Neurons Simulated Valves

```
real= [0.15 0.47 0.17; 0.42 0.23 0.74; 0.41 0.58 0.2; 0.09 0.51 1; 1 0.48 0.39; 1 0.37
      1; 1 0.53 0.2; 0.82 0.56 1;0.17 0.5 0.43];
E=transpose(real)
p = [1 0.5 0.2]; !input captured or feed in from field database
P=transpose(p); !correct
t = [0.93 0.35 0.71; 0.93 0.35 0.71; 0.93 0.35 0.71; 0.93 0.35 0.71; 0.93 0.35 0.71;
      0.93 0.35 0.71;0.93 0.35 0.71;0.93 0.35 0.71;0.93 0.35 0.71];
T=transpose(t); !the most only 2 different set  binary targets can be tracked accurately
net = newff([0 1;0 1;0 1],[1 3],{'logsig' 'logsig'});
net.trainParam.show = 50; !only 1set of  real target can be tracked accurately
net.trainParam.lr = 0.05;
net.trainParam.epochs = 38000;
net.trainParam.goal = 1e-24;
net.trainParam.gradient = 1e-15;
net.iw{1,1}=[0.3 0.5 0.1] ; !initial data
net.lw{2,1}=[0.1; 1; 0]; !Initialised data
net.b{1,1}=[0];      ! biased at '0' to reach best performance goal
net.b{2,1}=[0; 0; 0]
[net,tr]=train(net,E,T);
iw=net.iw{1,1}
lw=net.lw{2,1}
bias=net.b{1,1}
bias2=net.b{2,1}
Y = sim(net,P)
Tar=[0.93 0.35 0.71];
Target=transpose(Tar)
input=[1];
plot(input,Target,'r+',input,Y,'ko')
hold; xlabel('Input Sample');
ylabel('Output Y'); title('Output Layer Neurons'); hold;
```

Output Results : Output Y matches the target, with a given input P

```
TRAINLM
Epoch 0/38000, MSE 0.101745/1e-024, Gradient 1.43031/1e-010
Epoch 6/38000, MSE 1.77237e-025/1e-024, Gradient 1.78259e-014/1e-010
Performance goal met.
```

iw = 1.0e − 009 * − 0.0013
$\qquad$ 1.0e − 009 *0.2724
$\qquad$ 1.0e − 009 * 0.0274

lw = 0.9733
$\qquad$ 0.3275
$\qquad$ 0.2445

bias = 0.0014

bias2 = 2.0997
$\qquad$ − 0.7829
$\qquad$ 0.7731

Y =   0.9300
$\qquad$ 0.3500
$\qquad$ 0.7100

Target = 0.9300
$\qquad$ 0.3500
$\qquad$ 0.7100

9.11.3   Intelligent Advisor

We can establish on-line purity assessment and track the chemical reactions of the plant. Equipment from 'Magritek' can be used to monitor the state of the plant. Spinsolve software can be customized to track the plant reactions. Moreover, we can program a neural network system to advise on the differences between the actual chemical plant and the desired target quality (see reference [36] DeltaV system). The plant feedback variables from the automation, system or the chemical software are stored and retrieved by the computerised advisor. We can return position variables from the field valves into the computer neural network advisor by using ODBC. So, all the field instrumentation information, including the valve variables are updated continuously and stored in the database for retrieval.

Fig. 9.26 Quality matching advisor

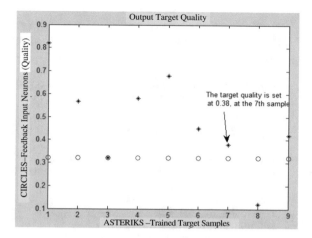

The NN advisor performs training for eight sampling targets for the chemical desired qualities. The neural network for the advisor uses 3-5-1 NN configuration. Each desired target shows in asterisk in the figure. For example, we select a target quality (target sample 7) at 0.38. So we feed in reference valves input [1 0.83 0.2] as the three valves variables to produce the target quality. The NN plant control system (P9.1) operates, and control its valves to their positions in the field. If the catalyst valve is stucked at position 0.89 (should be 1), the quality advisor will receive the mismatch target at 0.32 (target sample 3). Thus, the incorrect or not matching target lead us to troubleshoot the plant or valve error at the field site. Hence, the intelligent advisor system determines the site hardware failure to activate an alarm. Following up, we implement the fault correction at the field site (Fig. 9.26).

Program 9.2: Computerized NN Quality Advisor

```
real= [0.15 0.47 0.17; 0.42 0.23 0.74; 0.41 0.58 0.2; 0.09 0.51 1; 1 0.48 0.39; 1 0.37
      1; 1 0.83 0.2; 0.82 0.56 1;0.17 0.5 0.43];

E=transpose(real)
t = [0.8204 ;0.5678 ;0.32 ;0.58 ;0.679 ;0.45 ;0.38 ;0.12 ; 0.42];
T=transpose(t); !Target qualities trained
net = newff([0 1;0 1;0 1],[5 1],{'logsig' 'logsig'});
net.trainParam.show = 50; !3-5-1 configuration to train more targets
net.trainParam.lr = 0.05; !at a time.
net.trainParam.epochs = 38000;
net.trainParam.goal = 1e-24;
net.trainParam.gradient = 1e-15;
net.iw{1,1}=[0.3 0.5 0.1;0.3 0.5 0.1;0.3 0.5 0.1;0.3 0.5 0.1;0.3 0.5 0.1]
net.lw{2,1}=[0 0 0 0 0]; !initialisztions
net.b{1,1}=[0; 0; 0; 0; 0]
net.b{2,1}=[0]
W1 = net.iw{1,1}
W2 = net.lw{2,1}
b1= net.b{1,1}
b2= net.b{2,1}
[net,tr]=train(net,E,T);  !trainings
iw=net.iw{1,1}
lw=net.lw{2,1}
bias=net.b{1,1}
bias2=net.b{2,1}
Y1 = sim(net,E)
T
conne=database('db1',' ',' '); ! on-line updated database stored in db1
bo = exec(conne,'select all Field1 from Table1');
sor=fetch(bo,3);
a=sor.Data{1,1};
b=sor.Data{2,1};
c=sor.Data{3,1};
inp=[a;b;c]
! inp = [0.89; 0.83 ;0.2]; incoming 3 valves feedback retrieved from db1
Y = sim(net,inp); !simulated quality feedback from the field
iw=net.iw{1,1}
lw=net.lw{2,1}
bias=net.b{1,1}
bias2=net.b{2,1}
input=[1 2 3 4 5 6 7 8 9]; figure(2);
plot(input,Y1,'k*')
hold; xlabel('ASTERIKS - Trained Target Sample');
```

```
ylabel('CIRCLES - Feedback Input Neurons'); title('Output Target Quality');
plot(input,Y,'bo');
hold;
gtext('The target quality is set')
gtext('at 0.38, at the 7th sample')
```

Chapter 10
Computer Vision

Computer Vision is popular in the present automation, and control applications in the modern industries. Manufacturing and assembly lines made use of the technology to track and scan for product defects in the production lines. Cameras are fixed at the assembly lines to scan for bar code presence and other product parameters in the processed images.

10.1 Image Thresholding

In computer vision, we often encounter problems of errors when all the pixels has been thresholded. We often encounter two types of errors as a result of the extremes of the two final level classifications as either foreground or background.

1. We are not able to catch all the included pixels in the group.
2. Some pixels caught should not be in the group.

So the choice of thresholding comes in, to balance these two types of error. The threshold level is between 0 and 255 with reference to the black or the white background respectively. We implement a two steps thinning algorithm without violating the following constraints.

1. Does not remove end points.
2. Does not break connectivity.
3. Does not cause excessive erosion of the region.

© Springer Science+Business Media Singapore 2016
T.S. Ng, *Real Time Control Engineering*,
Studies in Systems, Decision and Control 65,
DOI 10.1007/978-981-10-1509-0_10

10.2   Zhang-Suen Thinning Algorithm

The mark-and-delete thinning algorithm has a template. The template is of size 3×3. It moves to overlap the image to determine the center pixel. It is an iterative algorithm and continues until no more black seeds removed. We used two sub-iterations in the algorithm.

First-step sub-iteration:

1. Connectivity (number of background-to-foreground, or '255' to '0' in this case) in template boundary pixel is one.
2. The foreground or '0' neighbouring pixels of the selected center template is within 2–6.
3. Either one of the pixels P(y, x − 1) P(y + 1, x) or P(y, x + 1) must be a white background.
4. Either one of the pixels P(y + 1, x) P(y, x + 1) or P(y − 1, x) must be a white (255) background.

Second-step sub-iteration:

First two procedures are the same as the first-step sub-iteration. The third and fourth steps are as follow:

3. Either one of the pixels P(y − 1, x) P(y, x + 1) or P(y, x − 1) must be a white background or value of 255.
4. Either one of the pixels P(y − 1, x) P(y + 1, x) or P(y, x − 1) must be a white background.

We flagged the seeds or pixels that satisfy the above conditions for changing. The flagged pixels changed after each step. The two-step process repeats until we can no longer change any more black pixels.

10.3   Brief Descriptions of the Program Algorithms

The program consists of two separate programs. Mainly the binary .m file and the thinning .m file. The binary program includes the 'Binary' function and the 'Binarise' sub-function. The main function process the seven different kinds of the images to read in and the different threshold level. It will call the binaries sub-function to do the binary processing. The outcome of this program will create seven images stored into different filenames each. The outcome is the binary images produced.

The second program 'thinning .m' which has the thinning function to do the thinning process. Besides, it also has six different sub-function to support the main program function. They are mainly the function: initialize, change, searchblack, transition, steponecd and steptwocd. The initialize function is to initialize the replacement data "AData" and the stored data "SData". After which the change function is to convert the black seeds or pixels into white pixels in the image. It is the thinning process. The 'searchblack' function is to do the neighbouring search for the black seeds on the required centered pixel. It searches for the eight surrounding pixels to find the black seeds. If the black pixels neighbour is between 2 and 6, then it activates a one for recognition in the main process of the program. The transition function finds a zero to one transition of the neighbouring pixels. There must be one transition only, to initiate for processing. The next function is to test for step 1c and 1d of the thinning algorithm. The last function 'steponecd' is to condition for the step 2, c and d conditions to meet its requirements. For the last two functions, a one will be activated if they meet the requirements. So altogether, the four ones will go through a judging process, whereby only three of the ones will be combined to check for pixels replacement into white seeds. Each of Zhang Suen's two step algorithm will test for three criteria, which will result in producing a 3 when we meet the four-condition of each step. This number of three will activate for conversion of the black seeds into white seeds.

In the Matlab thinning main program provided, the four conditions in each of the two steps are developed such that it only consists of three criteria in judging for removal. We will test each of the first and second conditions by going through a function subroutine. While we combine the third and fourth conditions to form criteria in conditioning for pixels removal. We apply the same to the third and fourth conditions in the second step, where only a function subroutine is used to test for the two conditions. Therefore, only 3 test functions criteria are needed to be satisfied in each step for pixel removal. The thinning algorithm consists of four conditions of which we mix the last two conditions to form criteria. The main algorithm for thinning is such that it loops through the two steps.

The program started off by reading into the program the image to thin. We take into account the size of the picture. As the width and the height, of the picture corresponds to the x, and y, which is the columns and rows of the image pixels concerned. Next, the main program of the thinning function initialized the matrices to be used. The thinning routine runs through pixel by pixel to search for the black seeds or pixels. Once we find it, it will be tested in the 1st step of the thinning algorithm. If it meets the three criteria of the first step, the test pixel or black seed will be flagged for replacement.

If not, the black seeds will be stored in another matrix location for testing during the second step of the algorithm. When all the pixels of the image are scan, the program will go to the change function to get the flagged seeds to be replaced by the white seeds to become the white background. Next, the remaining stored black pixels will be tested by the second step algorithm for replacement with the white seeds. It will go to the same procedure for pixels replacement once we flagged it. Otherwise, the remaining black seeds will be stored again for testing in the next round or iteration. The program repeats the two-step algorithm after the first two steps are over. It will search for the remaining stored black seeds for replacement with the white seeds. If no black pixels are to flag for replacement, then the program comes to an end. The final output becomes the thinned or skeleton (media axis) of the original image. We have to name the final image to store the final result. Zhang Suen's thinning algorithm thus provides the same outcome as the media axis transformation technique in gray level. It is most suitable for using as a skeleton transforming method in binary level.

```
-------------------------------------------
Program 10.1 Binarise Algorithm
-------------------------------------------
%Binarise Original Images
function Binary
global map imga h CData Y height width t Y1 image
image='img1.bmp' %define image
[imga,map]=imread(image); %input image
height=336;  %image height
width=339;  %image width
t=150; %threshold level
figure(1)  %open figure to view ready image
Binarise  %apply thresholding to view the character clearly
imwrite(Y1,map,'bimg1.bmp'); %save image as a filename

image='img3.bmp'  %an image is chosen
[imga,map]=imread(image);
height=349;
width=357;
t=70;  %threshold level is input defined (changeable)
figure(3)
Binarise
imwrite(Y1,map,'bimg3.bmp'); %save as a different filename

image='img4.bmp'
[imga,map]=imread(image);
height=746;
width=668;
t=70;
figure(4)
Binarise
imwrite(Y1,map,'bimg4.bmp');

image='img5.bmp'
[imga,map]=imread(image);
height=535;
width=576;
t=200;
figure(5)
Binarise
imwrite(Y1,map,'bimg5.bmp');

image='img6.bmp'
[imga,map]=imread(image);
height=436;
width=469;
t=70;
figure(6)
Binarise
imwrite(Y1,map,'bimg6.bmp');

function Binarise  %Binary function
global imga map Y CData h height width t Y1
```

```
h=image(imga);
colormap(map);
Y=get(h,'CData'); %grap image data (CData)
CData(1:height,1:width)=Y
for i=1:height %thresholding
   for j=1:width
if CData(i,j)>=t
  CData(i,j)=255;
elseif CData(i,j)<t
  CData(i,j)=0;
end
   end
end
Y1=CData(1:height,1:width) %Binarised matrix value
imwrite(Y1,map,'bimg.bmp'); %store temporary image in file
[Y1,map]=imread('bimg.bmp'); %read binarised image
imshow bimg.bmp; %show image
```

```
----------------------------------------==-
Program 10.2 Thinning Algorithm
----------------------------------------==---
function Thinning %thinning algorithm
global width height x y CData Yes AData seed SData
height=336; %image height depending on image input
width=339; %image width depending on image input
Yes=0; %initialisation
seed=0;
convert=0; %initialise convert
initialise; %initialise matrix
[Y1, map]=imread('bing1.bmp'); %changeable input image
h=image(Y1);
colormap(map);
Y=get(h,'CData');
CData(1:height,1:width)=Y;
for y=1:height %Perform Step1
   for x=1:width
     if CData(y,x)==0 %if black seed found
        searchblack; %do the 3 criterior of step 1
     out1=Yes; %or the 4 conditions of step 1
     transition;
     out2=Yes;
   steponecd;
   out3=Yes;
   out1+out2+out3;
   if out1+out2+out3==3 %if satisfy the 3 criterior to convert
   convert=1; %activate convert
   AData(y,x)=CData(y,x); %flagged for conversion
   else
   SData(y,x)=CData(y,x); %store black seed for next round thinning decision
```

```
      seed=seed+1; %indicate seed for thinning decision
      end
    end
end
end
change;  %convert black seed into white
for y=1:height  %Perform Step 2
  for x=1:width
    if (SData(y,x)==CData(y,x)) %if the stored black seed is found
    searchblack; % do the 3 criterior of step 2 again
    out1=Yes;    % or the 4 conditions of step 2 again
    transition;
    out2=Yes;
    steptwocd;
    out3=Yes;
    out1+out2+out3;
    if out1+out2+out3==3 % judge the 3 criterior
    convert=1; %activate convert
    AData(y,x)=CData(y,x);
    seed=seed-1;  %indicate seed left to be thinned
    else
    SData(y,x)=CData(y,x);%else store the remaining black seed for next round
    end
    end
  end
end
change;
while convert>0 %loop until no black seed is to be removed
convert=0;  %reinitialise convert
for y=1:height   %Perform Step 1
for x=1:width
  if SData(y,x)==CData(y,x) %if black seed found
    searchblack;
    out1=Yes;
    transition;
    out2=Yes;
    steponecd;
    out3=Yes;
    out1+out2+out3;
    if out1+out2+out3==3
    convert=1; %activate convert
    AData(y,x)=CData(y,x);
    seed=seed-1;
    else
    SData(y,x)=CData(y,x);
    end
  end
end
end
change;
for y=1:height  %Perform Step 2
```

```
        for x=1:width
          if SData(y,x)==CData(y,x) % if black seed found
            searchblack;
            out1=Yes;
            transition;
            out2=Yes;
            steptwocd;
            out3=Yes;
            out1+out2+out3;
            if out1+out2+out3==3
            convert=1; %activate convert
            AData(y,x)=CData(y,x);
            seed=seed-1;
            else
            SData(y,x)=CData(y,x);
            convert=0;  %check for no convert, may be removed for total thinning
            seed
            end
          end
        end
      end
      change;
      end
      Y1=CData(1:height,1:width) %Binarise matrix value
      imwrite(Y1,map,'timg11.bmp'); % store image into a file
      [Y1,map]=imread('timg11.bmp');  %readin image to be displayed
      figure(2);
      imshow timg11.bmp; %display image

      function initialise  %initialise AData & SData
      global x y height width AData SData
      for y=1:height
        for x=1:width
        AData(y,x)=150;  %initialise to intermediate value
        SData(y,x)=150;  %of between black and white
        end
      end

      function change  %black seed change into white
      global CData AData x y height width
      for y=1:height
        for x=1:width
        if AData(y,x)==CData(y,x)
        CData(y,x)=255;
        end
      end
      end

      function searchblack  %finding black neighbours
      global CData x y Yes
      b=0;
```

```
Yes=0;
if CData(y-1,x-1)==0
  b=b+1;  end
if CData(y+1,x+1)==0
  b=b+1; end
if CData(y+1,x)==0
  b=b+1; end
if CData(y-1,x)==0
  b=b+1; end
if CData(y,x+1)==0
  b=b+1; end
if CData(y,x-1)==0
  b=b+1; end
if CData(y+1,x-1)==0
  b=b+1; end
if CData(y-1,x+1)==0
  b=b+1; end
b;
if (b>=2)
  if (b<=6)
   Yes=1; end
end

function transition  %zero to one transition
global Yes CData x y
rt=0;
if (CData(y,x-1)>0)
  if(CData(y+1,x-1)<255)
   rt=rt+1; end
end
if (CData(y+1,x-1)>0)
  if (CData(y+1,x)<255)
   rt=rt+1;  end
end
if (CData(y+1,x)>0)
  if (CData(y+1,x+1)<255)
   rt=rt+1;  end
end
if (CData(y+1,x+1)>0)
  if (CData(y,x+1)<255)
   rt=rt+1;  end
end
if (CData(y,x+1)>0)
  if (CData(y-1,x+1)<255)
   rt=rt+1;  end
end
if (CData(y-1,x+1)>0)
  if (CData(y-1,x)<255)
   rt=rt+1;  end
end
if CData(y-1,x)>0
  if CData(y-1,x-1)<255
   rt=rt+1;  end
end
if CData(y-1,x-1)>0
  if CData(y,x-1)<255
   rt=rt+1;  end
end
rt;
if rt==1
  Yes=1;
else
  Yes=0;
end

function steponecd    %Step 1c and 1d conditions
global Yes CData x y
Yes=0;
if (CData(y,x-1)+CData(y+1,x)+CData(y,x+1))>0
  if(CData(y+1,x)+CData(y,x+1)+CData(y-1,x))>0
   Yes=1;
   end
end

function steptwocd %step 2c and 2d conditions
global CData x y Yes
Yes=0;
if (CData(y,x-1)+CData(y-1,x)+CData(y,x+1))>0
  if (CData(y,x-1)+CData(y-1,x)+CData(y+1,x))>0
   Yes=1;
  end
end
```

10.4   Image Results

In the first five sets of the character examples, pictures two, three, and five, give the same amount of thresholding of about 70 into the image, while that of image 1, thresholds at 150 and image four thresholds at a value of 200. For image one, when we

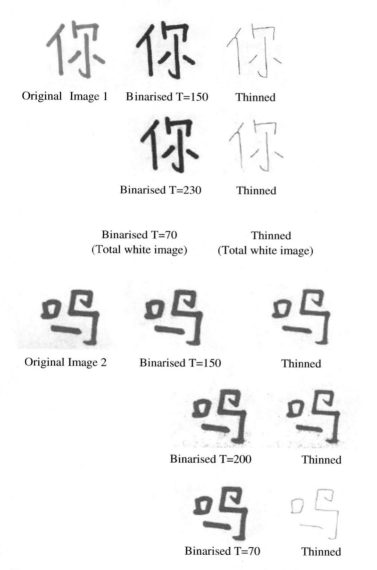

Original Image 1   Binarised T=150   Thinned

Binarised T=230   Thinned

Binarised T=70   Thinned
(Total white image)   (Total white image)

Original Image 2   Binarised T=150   Thinned

Binarised T=200   Thinned

Binarised T=70   Thinned

Fig. 10.1 Chinese characters (images sets 1 and 2). *Note* T represents threshold level within 0–255

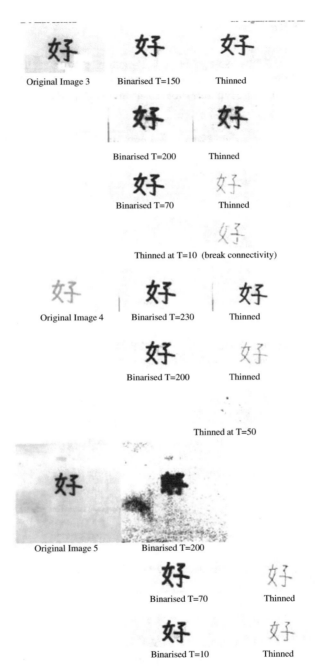

Fig. 10.2 More Chinese characters (images sets 3–5)

input a binarized level of 150 will perform the thinning well. If a value of 230 is input, the resulting character after thinning will still be fined. But if we apply a threshold level of 70, it will result in a blank image after binarized, so is after thinning.

Image 2 is binarized at a value of 70. We can binarize the image at a level of 10 also. The outcome after thinning is as good. However, if we use a value of 200 or 150 as the threshold level, the binarized image will produce some black patches. And if thinned, the character will not be a skeleton. It will still be fat with only minor trimmings of the character.

The black and dark original image 3, if binarized at 200, will cause the creation of the black border boundary, which is far away from the character, to appear in the image after thinning. It is the result of the excessive level of the threshold value applied. However, when thresholded at 150, the final thinning result will still create a slightly fatter image character. It is not the full thinning operation, and we cannot produce the skeleton outcome. On the other hand, when binarized at a value of about 10, will result in the breaking connectivity of the character when after thinning. Image 4 has a faint image, so a higher value of the thresholding is needed. Image 4 erodes when thresholded at a level of 50. The final result of the eroded character after thinning will be a blank white image without any character in view. If binarized at a value of 230, or slightly over thresholded, the outcome will be a slightly fatter skeleton, which will still look like a character, but only never skeletoned. Moreover, the borderline of the image will appear.

In image 5, the average threshold is at 70–10. It is because it has a dark original image background, so lesser thresholding level is required. However, if we binarized at a very large value of 200, the character itself will be over darken and tends to become overconnected as if black ink leaks out from the character. Besides, black dots started to appear in the image from nowhere. The character will worsen when we thinned it. It will become a dark patch of an unrecognised character (Figs. 10.1, 10.2).

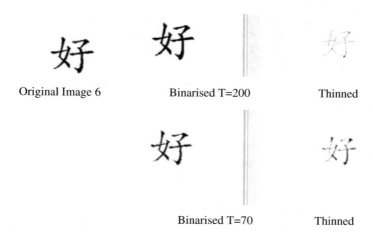

| Original Image 6 | Binarised T=200 | Thinned |
| Binarised T=70 | Thinned |

Fig. 10.3 Hao Chinese character (image set 6)

Original Image 7 (8-bit depth bitmap)

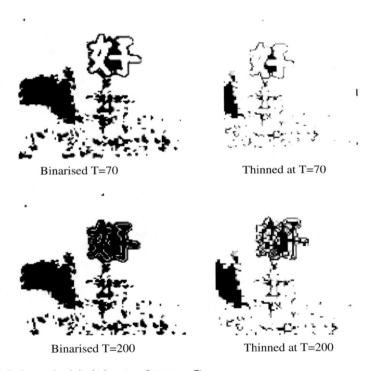

Binarised T=70   Thinned at T=70

Binarised T=200   Thinned at T=200

Fig. 10.4 Scanned original character (image set 7)

So thresholding adjustments are for improving the appearance and visibility of the characters at the binary level, which will in turn affect the thinning result of the characters. Let us look at the image set 6. The set of image thinned lesser at a lower threshold of 70 than at threshold level of 200. We can see that it has the reverse thinning effect from the rest of the image sets. Only this original image set is exceptional. Moreover, it is noted that the scanned image (image set 7) after binarization, cannot produce the proper characters for thinning. It is because the given image is only an 8-bit colour depth image. It represents only one (8-bit) out of the three primary colour (24-bit) components. Zhang Suen's method of thinning provides an erosion of the original image into a skeleton. Likewise, it can also just thinned a bit, which results in the thick character still remainings. The thresholding level affects the thinnings of the characters. If the threshold level is too high towards white, applying thinning algorithm will convert more to the black side which will eventually darken the image. However, if the threshold level is small towards the value of black (zero), applying thinning will convert lesser of the value to the black region thereby causing connectivity brokage of the character. It, therefore, results in the eroding of the character from the original image. That is what happens to the final results to some of the characters applied (Figs. 10.3, 10.4).

Appendix A
MC68HC11 Registers

```
def PORTA     1000 u8
def DDRA      1001 u8  # 68HC11F1...
def PIOC      1002 u8
def PORTG     1002 u8  # 68HC11F1...
def DDRG      1003 u8  # 68HC11F1...
def PORTC     1003 u8
def PORTB     1004 u8
def PORTCL    1005 u8
def PORTF     1005 u8  # 68HC11F1...
def PORTCF1   1006 u8  # 68HC11F1...
def DDRC      1007 u8
def PORTD     1008 u8
def DDRD      1009 u8
def PORTE     100A u8
def CFORC     100B u8
def OC1M      100C u8
def OC1D      100D u8
def TCNT      100E u16
def TIC1      1010 u16
def TIC2      1012 u16
def TIC3      1014 u16
def TOC1      1016 u16
def TOC2      1018 u16
def TOC3      101A u16
def TOC4      101C u16
def TOC5      101E u16
def TCTL1     1020 u8
def TCTL2     1021 u8
def TMSK1     1022 u8
def TFLG1     1023 u8
def TMSK2     1024 u8
def TFLG2     1025 u8
def PACTL     1026 u8
def PACNT     1027 u8
def SPCR      1028 u8
def SPSR      1029 u8

def SPDR      102A u8
def BAUD      102B u8
def SCCR1     102C u8
def SCCR2     102D u8
def SCSR      102E u8
def SCDR      102F u8
def ADCTL     1030 u8
def ADR1      1031 u8
def ADR2      1032 u8
def ADR3      1033 u8
def ADR4      1034 u8
def OPT2      1038 u8  # 68HC11F1...
def OPTION    1039 u8
def COPRST    103A u8
def PPROG     103B u8
def HPRIO     103C u8
def INIT      103D u8
def TEST1     103E u8
def CONFIG    103F u8
def CSSTRH    105C u8  # 68HC11F1...
def CSCTL     105D u8  # 68HC11F1...
def CSGADR    105E u8  # 68HC11F1...
def CSGSIZ    105F u8  # 68HC11F1...
def SCONF     0420 u8  # X68C75
def PORTBO    0410 u8  # X68C75
def PORTBI    0430 u8  # X68C75
def PORTCO    0408 u8  # X68C75
def PORTCI    0428 u8  # X68C75
```

© Springer Science+Business Media Singapore 2016
T.S. Ng, *Real Time Control Engineering*,
Studies in Systems, Decision and Control 65,
DOI 10.1007/978-981-10-1509-0

```
'        startup.bas
ProgramPointer  $8000
DataPointer     $2000
StackPointer    $7FFF


        sect    text
        cli                     ; enable debugger
        ldx     #_data_s
        bra     _crt2
_crt1   clr     0,x             ; clear data area
        inx
_crt2   cpx     #_data_e
        bne     _crt1
        sect    data
_data_s         equ *

byte PORTA   at $1000           ' mode single-chip
byte DDRA    at $1001
byte PORTG   at $1002
byte DDRG    at $1003
byte PORTF   at $1005
byte PORTD   at $1008
byte DDRD    at $1009
byte PORTE   at $100A
int  TCNT    at $100E
byte TCNTL   at $100F
byte TMSK2   at $1024
byte TFLG2   at $1025
byte PACTL   at $1026
byte PACNT   at $1027
byte BAUD    at $102B
byte SCCR1   at $102C
byte SCCR2   at $102D
byte SCSR    at $102E
byte SCDR    at $102F
byte ADCTL   at $1030
byte ADR     at $1031
byte OPTIONS at $1039
byte PORTB   at $1060
byte PORTC   at $1061
byte PORTM   at $1062
byte PORTN   at $1063
```

Appendix B
MCU Port Testers

© Springer Science+Business Media Singapore 2016
T.S. Ng, *Real Time Control Engineering*,
Studies in Systems, Decision and Control 65,
DOI 10.1007/978-981-10-1509-0

```
porta=porta and $df
delay(10)
porta=porta or $ff
delay(10)
porta=porta and $bf
delay(10)
```

'PORTA TESTER

```
ProgramPointer  $2000
DataPointer     $0002
StackPointer    $7FEF

byte ddra at $1001
byte porta at $1000
int q,s
ddra=$ff
porta=$ff
ASM cli
do
q=0;s=0
for q=1 to 3
porta =porta or $ff
delay(10)
porta=porta and $fe
delay(10)
porta=porta or $ff
delay(10)
porta=porta and $fd
delay(10)
porta=porta or $ff
delay(10)
porta=porta and $fb
delay(10)
porta=porta or $ff
delay(10)
porta=porta and $f7
delay(10)
porta=porta or $ff
delay(10)
porta=porta and $ef
delay(10)
porta=porta or $ff
delay(10)
```

```
porta=porta or $ff
delay(10)
porta=porta and $7f
delay(10)
next q
porta=porta or $ff
for s=1 to 10
porta=porta xor $ff
delay(100)
next s
loop

function delay(c)
int i
for c=c to 0 step -1
for i= 0 to 1000
next i
next c
end function


"PORTB TESTER

'byte ddrb at $1001
byte porta at $1004
int q,s
porta=$ff
ASM cli
do
porta=$55
delay(100)
porta=$aa
delay(100)
loop
```

```
function delay(c)
int i
for c=c to 0 step -1
for i= 0 to 1000
next i
next c
end function
```

'PORTC TESTER

```
int j,f
byte portg        at $1002
byte ddrg         at $1003
ddrg.1 = 1
byte portc at $1061 'no need
define ddrc
'byte title()=" Microprocessor"
'byte portc at $1006 'single chip
mode no need define ddrc
for j=100 to 0 step -1
for f= 0 to 1000
portg.1=0
next f
next j
ASM cli
do
if portc.4=0 then portg.1=0
if portc.4=1 then portg.1=1
portc.4=0 'useless command
loop

function delay(c)
int i
for c=c to 0 step -1
for i= 0 to 1000
next i
next c
end function
```

'PORTD TESTER

```
byte portd at $1008
byte ddrd at $1009
byte portg at $1002
byte ddrg at $1003

ddrd=$3f
ddrg=$0c
portd=$f0
ASM cli


do
portg=$07
delay(300)
portg=$0b
delay(300)
portg=$ff
portd=$df
delay(300)
portd=$ef
delay(300)
portd=$f7
delay(300)
portd=$fb
delay(300)
portd=$ff
loop

function delay(c)
int i
for c=c to 0 step -1
for i= 0 to 1000
next i
next c
end function
```

'PORTE TESTER

```
ProgramPointer  $8000
DataPointer     $0002
StackPointer    $7FEF
```

```
byte porte at $100A
byte ADCTL at $1030
byte ADR at $1031
byte OPTIONS at $1039
ADR=$00
OPTIONS=$90
delay(5)
ASM cli
do
analogin(2) 'BIT NO. 7
loop

function delay(c)
int i
for c=c to 0 step -1
for i= 0 to 1000
next i
next c
end function

function analogin(ch)
ADCTL = ch
do
loop until ADCTL.2=1
return ADR
end function

function read(value)
int cha, old_cha
if cha==old_cha then return 0
old_cha = cha
return cha
end function


'PORTF TESTER

ProgramPointer  $8000
DataPointer     $2000
StackPointer    $7FEF
byte portf at $1005
byte portc0 at $0408
byte portn at $1063
byte portm at $1062
```

```
byte portg at $1002
byte ddrg at $1003
ddrg.1=1
ASM cli
do
'direct debug at $1063
'read(0)
'portn.7=0 'useless command
analogout(4, 0xff)
if portn.4=0 then portg.1=0
if portn.4=1 then portg.1=1
loop

function analogout(ch, val)
' 1:A, 2:B, 4:C
portm=0x20  'cs=0
max512(ch)
max512(val)
portm=0x30  'cs=1
end function

function max512(val)
byte cnt
for cnt=0 to 7
if val and 0x80 then
portm=0x22  'sdin=1
portm=0x23  'sclk=1
else
portm=0x20  'sdin=0
portm=0x21  'sclk=1
end if
portm=0x20  'sclk=0
val=val+val  'shift left
next cnt
end function

function delay(c)
int i
for c=c to 0 step -1
for i=0 to 1000
next i
next c
end function
```

Appendix C
LCD References

See Tables A.1 and A.2.

Table A.1 Control and display command

KS0070B   16COM/80SEG DRIVER & CONTROLLER FOR DOT MATRIX LCD

CONTROL and DISPLAY COMMAND(continued)

Command	RS	R/W	DB₇	DB₆	DB₅	DB₄	DB₃	DB₂	DB₁	DB₀	Excution time (fosc=250KHz)	Remark
SET CG RAM ADDRESS	L	L	L	H	CG RAM address (corresponds to cursor address)						42μs	CG RAM Data is sent and received after this setting
SET DD RAM ADDRESS	L	L	H	DD RAM address							42μs	DD RAM Data is sent and received after this setting
READ BUSY FLAG & ADDRESS	L	H	BF	Address Counter used for Both DD & CG RAM address							0μs	BF H Busy / L Ready - Reads BF indication internal operating is being performed. - reads address counter contents
WRITE DATA	H	L	Read Data								46	Write data into DD or CGRAM
READ DATA	H	H	Write Data								46μs	Read data from DD or CGRAM

X: Don't care
Table 1

© Springer Science+Business Media Singapore 2016
T.S. Ng, *Real Time Control Engineering*,
Studies in Systems, Decision and Control 65,
DOI 10.1007/978-981-10-1509-0

Table A.2 The command control codes

Command	Binary								He*
	D7	D6	D5	D4	D3	D2	D1	D0	
Clear display	0	0	0	0	0	0	0	1	01
Display and cursor home	0	0	0	0	0	0	1	x	02 or 03
Character entry mode	0	0	0	0	0	1	1/D	S	04 to 07
Display on/off and cursor	0	0	0	0	1	0	U	B	08 to 0F
Display/cursor shift	0	0	0	1	D/c	R/L	X	x	10 to IF
Function set	0	0	1	8/4	2/1	10/7	X	x	20 to 3F
Set CGRAM address	0	1	A	A	A	A	A	A	40 to 7F
Set display address	1	A	A	A	A	A	A	A	80 to FF

I/D: 1 = increment*, 0 = decrement
S: 1 = Display shift on. 0 = Display shift off*
O; 1 = Display on, 0 = Display off*
U: 1 = Cursor underline on, 0 = Underline off*
B: 1 = Cursor blink on, 0 = Cursor blink off*
D/C: 1 = Display shift, 0 = Cursor move
R/L: 1 = Right shift, 0 = Left Shift
B/4: 1 = 8 bit interface*, 0 = 4 bit interface
2/1: 1 = 2 Iine mode, 0 = 1 line mode*
10/7: 1 = 5 × 10 dot format. 0 = 5 × 7 dot format*
x = Don't care
* = Initialisation settings

Char.code

	0000	0010	0011	0100	0101	0110	0111	1010	1011	1100	1101	1110	1111
xxxx0000			0	@	P	`	p		―	タ	ミ	α	p
xxxx0001		!	1	A	Q	a	q	。	ア	チ	ム	ä	q
xxxx0010		"	2	B	R	b	r	「	イ	ツ	メ	β	θ
xxxx0011		#	3	C	S	c	s	」	ウ	テ	モ	ε	∞
xxxx0100		$	4	D	T	d	t	、	エ	ト	ヤ	μ	Ω
xxxx0101		%	5	E	U	e	u	・	オ	ナ	ユ	σ	ü
xxxx0110		&	6	F	V	f	v	ヲ	カ	ニ	ヨ	ρ	Σ
xxxx0111		'	7	G	W	g	w	ア	キ	ヌ	ラ	q	π
xxxx1000		(	8	H	X	h	x	イ	ク	ネ	リ	√	x̄
xxxx1001		)	9	I	Y	i	y	ゥ	ケ	ノ	ル	⁻¹	y
xxxx1010		*	:	J	Z	j	z	エ	コ	ハ	レ	i	千
xxxx1011		+	;	K	[	k	{	ォ	サ	ヒ	ロ	x	万
xxxx1100		,	<	L	¥	l			ャ	シ	フ	ワ	¢	円
xxxx1101		-	=	M	]	m	}	ュ	ス	ヘ	ン	₤	÷
xxxx1110		.	>	N	^	n	→	ョ	セ	ホ	゛	ñ	
xxxx1111		/	?	O	_	o	←	ッ	ソ	マ	゜	ö	■

References

1. http://www.datasheet4u.com/datasheet/7/4/L/74LS08_FairchildSemiconductor.pdf.html (2-input AND gate)
2. http://www.datasheet4u.com/datasheet/7/4/L/74LS32_FairchildSemiconductor.pdf.html (2-input OR gate)
3. http://www.datasheet4u.com/datasheet/7/4/L/74LS04_FairchildSemiconductor.pdf.html (hex inverting gate)
4. http://html.alldatasheet.com/html-pdf/171559/TI/74LS244/24/1/74LS244.html (Buffer or Line driver)
5. http://www.ti.com/lit/ds/symlink/sn74ls74a.pdf (Dual D-Latch)
6. http://docs.google.com/viewer?url=http%3A%2F%2Fwww.datasheet.hk%2Fdownload_online.php%3Fid%3D1027398%26pdfid%3D384EA12E7F41ADD153FD14E6B124409A%26file%3D0021%5Csn74ls76_186750.pdf&embedded=true (Negative-edge triggered J-K flip-flop)
7. http://www.alldatasheet.com/datasheet-pdf/pdf/22437/STMICROELECTRONICS/L298.html (Motor driver)
8. http://www.pepperl-fuchs.com/global/en/classid_142.htm (proximity sensor)
9. http://ph.parker.com/sg/en/solenoid-valves (soleniod valve)
10. http://pdf.datasheetcatalog.com/datasheet/siemens/BUZ12.pdf (BUZ12)
11. http://www.datasheetarchive.com/dl/Datasheet-093/DSA0066289.pdf (IN4001 diode)
12. http://cache.freescale.com/files/microcontrollers/doc/data_sheet/M68
13. HC12B.pdf?pspll=1 (MCU 68HC12 family)
14. http://html.alldatasheet.com/html-pdf/50893/FAIRCHILD/7407/403/1/7407.html (open-collector output hex buffer)
15. http://www.datasheetarchive.com/dlmain/Databooks-2/Book259-13.pdf (4N28 opto-coupler)
16. http://www.alldatasheet.com/datasheet-pdf/pdf/8979/NSC/LM555.html (555 timer)
17. http://www.ti.com/lit/ds/symlink/uln2003a.pdf(darlington transistor)
18. http://courses.cs.tau.ac.il/embedded/docs/LPC2148_Education_Board/KS0070B.pdf (LCD dot matrix)
19. http://www.ee.nmt.edu/~rison/ee308_spr98/supp/feb_9/hc11_h.html (hc11.h file).
20. The Intel Microprocessors 8086/8088,..80186,486..pentium,..core2 Architecture, Programming & Interfacing - (2009) by Barry B. Brey
21. http://www.emsl.pnl.gov (process control)
22. Evolutionary Learning Algorithms for Neural Adaptive Control (Perspectives in Neural Computing) by Dimitris Dracopoulos (Sep 12, 1997)
23. http://www.neurodimension.com/
24. A neural network approach to on-line monitoring of machining processes by Raju G Khanchustambham, Neural Networks, 1992. IJCNN., International Joint Conference on (Volume:2),1992.

25. http://www.documentation.emersonprocess.com/groups/public/documents/specification_
    sheets/d301714x012.pdf (Remote Operation controller ROC800)
26. http://www.documentation.emersonprocess.com/groups/public/documents/specification_
    sheets/d301731x012.pdf (Network Radio Module NRM)
27. http://www2.emersonprocess.com/siteadmincenter/PM%20Rosemount%20Documents/008
    13-0100-4075.pdf (Smart Wireless THUM Adapter)
28. http://www2.emersonprocess.com/siteadmincenter/pm%20rosemount%20documents/00813-
    0100-4648.pdf (Rosemount 648 Wireless Temperature Transmitter)
29. http://www.documentation.emersonprocess.com/groups/public/documents/instruction_
    manuals/d103621x012.pdf (Wireless Valve Position Monitoring)
30. http://www.westlockcontrols.com/products/wireless/?id=tcm:528-34768&catid=tcm:528-32
    300-1024#product-description-tab (Wireless Valve Monitoring System)
31. http://westlockcontrols.com/Images/WESTDS-09082-EN-1304.pdf (Manual for Wireless
    Valve Monitoring System)
32. http://www2.emersonprocess.com/siteadmincenter/PM%20Articles/VM_FALL12_Wireless_
    Reprint.pdf (Wireless Valve Automation)
33. http://www.scientistlive.com/content/flexible-solution-high-throughput-evaporation?dm_i=
    371,3ZRZE,BIJ57N,EF9MF,1 (Evaporation Monitoring)
34. http://www.scientistlive.com/content/routine-purity-assessment-and-reaction-monitoring?
    dm_i=371,3ZRZE,BIJ57N,EF9MF,1 (Reaction Monitoring)
35. http://www.magritek.com/2015/12/17/magritek-introduce-reaction-monitoring-kits-for-
    spinsolve-benchtop-nmr/ (NMR Monitor)
36. www.EmersonProcess.com/DeltaV (DeltaV system)
37. http://www.julabo.com/ (Temperature control technology)

Index

© Springer Science+Business Media Singapore 2016
T.S. Ng, *Real Time Control Engineering*,
Studies in Systems, Decision and Control 65,
DOI 10.1007/978-981-10-1509-0

Printed in the United States
By Bookmasters